Reveal MATH®

Student Edition

Grade 1 • Volume 2

Mc
Graw
Hill

Back cover: Welsh Designs/Stockimo/Alamy Stock Photo

mheducation.com/prek-12

Send all inquiries to:
McGraw Hill
8787 Orion Place
Columbus, OH 43240

ISBN: 978-0-07-683904-9
MHID: 0-07-683904-4

Printed in the United States of America.

5 6 7 8 9 QSX 24 23 22

Contents in Brief

Welcome to *Reveal Math*!

We are excited that you have made us part of your math journey.

Throughout this school year, you will explore new concepts and develop new skills. You will expand your math thinking and problem-solving skills. You will be encouraged to persevere as you solve problems, working both on your own and with your classmates.

With *Reveal Math*, you will experience activities to spark your curiosity and challenge your thinking. In each lesson, you will engage in sense-making activities that will make you a better problem solver. You will have different learning experiences to help you build understanding.

We look forward to revealing to you the wonder and excitement of math.

The *Reveal Math* authors

The *Reveal Math* Authorship Team

McGraw-Hill teamed up with expert mathematicians to create a program centered around you, the student, to make sure each and every one of you can find joy and understanding in the math classroom.

Ralph Connelly, Ph.D.
Authority on the development of early mathematical understanding.

Annie Fetter
Advocate for students ideas and student thinking that fosters strong problem solvers.

Linda Gojak, M.Ed.
Expert in both theory and practice of strong mathematics instruction.

Sharon Griffin, Ph.D.
Champion for number sense and the achievement of all students.

Ruth Harbin Miles, Ed.S.
Leader in developing teachers' math content and strategy knowledge.

Susie Katt, M.Ed.
Advocate for the unique needs of our youngest mathematicians.

Nicki Newton, Ed.D.
Expert in bringing student-focused strategies and workshops into the classroom.

John SanGiovanni, M.Ed.
Leader in understanding the mathematics needs of students and teachers.

Raj Shah, Ph.D.
Expert in both theory and practice of strong mathematics instruction.

Jeff Shih, Ph.D.
Advocate for the importance of student knowledge.

Cheryl Tobey, M.Ed.
Facilitator of strategies that drive informed instructional decisions.

Dinah Zike, M.Ed.
Creator of learning tools that make connections through visual-kinesthetic techniques.

Meanings of Addition

Meanings of Subtraction

Addition within 100

Compare Using Addition and Subtraction

Unit 11

Subtraction within 100

Measurement and Data

Equal Shares

Let's Talk About Math!

This year, you will explore the language of mathematics together as you talk about math with your classmates. You are going to learn many new words. Use these resources as you expand your vocabulary.

Glossary

In the back of this book, you will find a glossary with definitions.

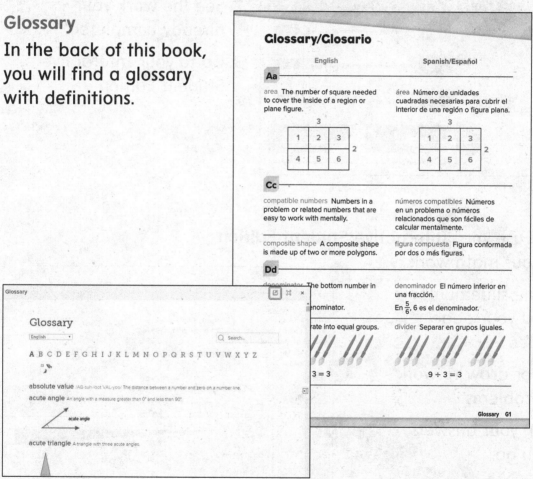

Interactive Glossary

The Interactive Glossary will support you as you work in your Interactive Student Edition.

Jump into Learning!

You can find all the resources you need from your **Student Dashboard**.

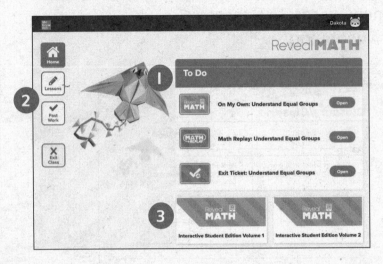

1. See your work in the To-Do List.

2. See the work you already completed.

3. Go to your Interactive Student Edition.

You can use your **Interactive Student Edition** for all your math work.

1. Use the slide numbers to find your page number.

2. Type or draw to work out problems.

3. Check your answers as you go.

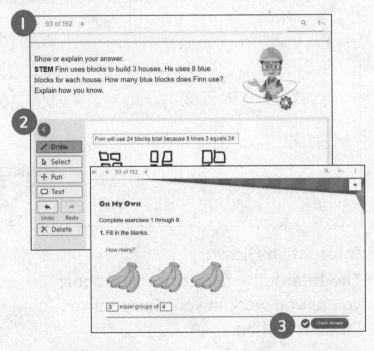

Access Lesson Supports Online!

You can also use these to support while you practice.

 | GO ONLINE

Need an Instant Replay of the Lesson Content?

Each lesson has a **Math Replay** video that provides a 1–2 minute overview of the lesson concept.

Virtual Tools to Help You Problem-Solve

You can access the eToolkit at any time from your Student Dashboard. You can access these tools:

- Counters
- Base-Ten Blocks
- Array Builder
- Fraction Model
- Bucket Balance

- Geometry Sketch
- Money
- Fact Triangles
- Number Line
- and more!

Key Concepts and Learning Objectives

Key Concept Habits of Mind and Classroom Norms
for Productive Math Learning

- I make sense of problems and think about numbers and quantities. (Unit I)

- I share my thinking with my classmates. (Unit I)

- I can use math to make sense of everyday problems. (Unit I)

- I see patterns in math. (Unit I)

- I describe my math story. (Unit I)

- I work productively with my classmates. (Unit I)

Key Concept Addition and Subtraction

- I relate counting to addition. (Unit 4)

- I relate counting to and counting back to subtraction. (Unit 5)

- I use different strategies to add and subtract within 100. (Units 4, 5, 9, 11)

- I use addition to solve problems involving adding to and putting together. (Unit 7, 8, 10)

- I explain what the equal sign means. (Unit 4)

Key Concept Number Sense and Place Value

- I read and write numbers from 0 to 120. (Unit 2)

- I use place value to represent 2-digit numbers. (Unit 3)

- I explain that 10 ones equal 1 ten. (Unit 3)

- I compare two 2-digit numbers by comparing the number of tens and the number of ones. (Unit 3)

Key Concept Measurement and Data

- I order three objects from shortest to longest. (Unit 12)

- I compare the lengths of two objects. (Unit 12)

- I measure the length of objects. (Unit 12)

- I tell time to the nearest hour and half hour. (Unit 12)

- I organize and interpret data into three categories. (Unit 12)

Key Concept Attributes of Shapes

- I can describe attributes that define shapes. Some defining attributes are the number of sides and the number of angles. (Unit 6)

- I can describe attributes that do not define shapes. Some non-defining attributes are color, size, and orientation. (Unit 6)

- I can compose two-dimensional and three-dimensional shapes to create composite shapes. (Unit 6)

- I can partition circles and rectangles into 2 and 4 equal parts. (Unit 13)

- I can describe the 2 and 4 equal parts of a circle or rectangle. (Unit 13)

Math is...

How would you complete this sentence?

Math is.....

Math is not just adding and subtracting.

Math is...
- working together
- finding patterns
- sharing ideas
- listening thoughtfully to our classmates
- sticking with a task even when it is a little challenging

In *Reveal Math,* you will develop the habits of mind that strong doers of math have. You will see that math is all around us.

Let's be Doers of Mathematics

Remember, math is more than getting the right answer. It is a tool for understanding the world around you. It is a language to communicate and collaborate. Be mindful of these prompts throughout the year to access the power of math.

1. **Math is... Mine**
   - Mindset

2. **Math is... Exploring and Thinking**
   - Planning
   - Connections
   - Thinking

3. **Math is... My World**
   - In My World
   - Modeling
   - Choosing Tools

4. **Math is... Explaining and Sharing**
   - Explaining
   - Sharing
   - Precision

5. **Math is... Finding Patterns**
   - Patterns
   - Generalizations

6. **Math is... Ours**
   - Mindset

Lesson 3-1
Understand Equal Groups

Be Curious

What do you notice?
What do you wonder?

Math is... Mindset
What can you do to be an active listener?

Math is... Mindset

What can you do to be an active listener?

Explore the Exciting World of STEM!

Ever wonder how math applies in the real world? In every unit, you will learn about a STEM career, from protecting our parks to exploring outer space. You will learn about the STEM career through digital simulations and projects.

STEM Career Kid: Meet Sienna
Let the STEM Career Kid introduce their career and talk about the different responsibilities.

Math In Action: Nutritionist
Watch the Math in Action to see how the math you are learning applies to the real world.

Hi, I'm Sienna.
I want to be a nutritionist to help people eat to feel great!

Meanings of Addition

Focus Question

How can I solve addition problems?

Hi, I'm Chloe.

I want to be a carpenter. Today I am making a treehouse. One board is 3 feet long. Another is 4 feet long. How long will they be if I put them together? I can add to find out!

Name

Perfect Triangle

Use each number 1–6 exactly once so that the sum of the numbers along each side of the triangle is the same.

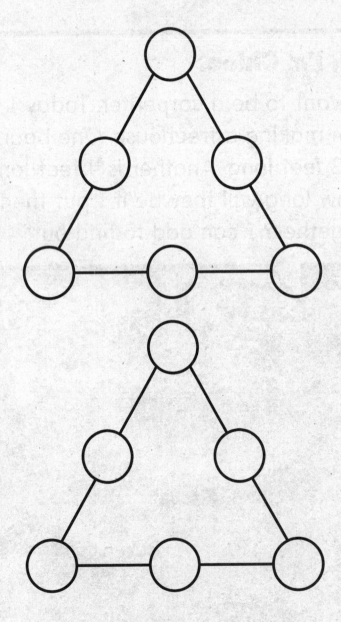

Represent and Solve Add To Problems

Be Curious

What question could you ask?

Math is... Mindset

What helps you be ready to learn?

Learn

9 birds are on a fence.
8 more birds land on
the fence.

**How many birds are
on the fence?**

You can use cubes to show the birds.

9 cubes   8 more cubes

An **equation** shows the problem.

$9 + 8 = ?$

Add the cubes.

$9 + 8 = 17$

17 birds are on the fence.

Math is... Modeling

How do cubes show
the birds?

You can add to find how many
in all.

Work Together

5 raindrops fall. 6 more raindrops fall.
How many raindrops fall?
Show your thinking.

_____ raindrops

On My Own

Name _____

1. Peter has 6 berries. He gets 7 more berries. How many berries does he have? Which equation matches the word problem? Circle the equation.

$7 - 6 = ?$ $6 + 7 = ?$

What equation can show the problem? Use ? for the unknown. Then solve.

2. There are 9 cars. 3 more cars drive up. How many cars are there?
   Equation:

   _____ cars

3. Nisha has 7 stickers. She gets 4 more stickers. How many stickers does Nisha have?
   Equation:

   _____ stickers

4. There are 3 birds in a tree. 14 more birds land in the tree. How many birds are in the tree?
   Equation:

   _____ birds

5. **STEM Connection** Chloe has 9 nails.
She gets 6 more nails. How many
nails does Chloe have now?
Draw to show your thinking.

_____ nails

6. **Extend Your Thinking** Make up a word problem.
Make the total unknown. Then solve.

Reflect

How do you know you need to add to solve
a word problem?

Math is... **Mindset**

How did you get ready
to learn today?

Represent and Solve More Add To Problems

?

Be Curious

Jay has some oranges.

How many more oranges does he need to have more oranges?

Math is... Mindset

How can you help identify a problem in your class or community?

Learn

Jay has 6 oranges. He gets some more oranges. Now he has 11 oranges.

How many oranges does he get?

6 counters	? more counters	11 counters

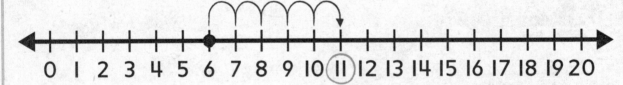

$6 + ? = 11$

Count on to find the missing addend.

$6 + 5 = 11$
Jay gets 5 oranges.

Math is... Choosing Tools

What other tool can you use to find the sum?

Work Together

Kathy has 7 peaches. She buys more peaches. Now she has 16 peaches. How many peaches did Kathy buy?

Show your thinking.

_____ peaches

On My Own

Name _____

1. Carlos has 7 shells. He finds more shells. Now he has 12 shells. How many shells did Carlos find? Make an equation to show the problem.
Use ? for the unknown.
Equation:

How can you make an equation to show the problem? Use ? for the unknown. Then solve.

2. Kyle has some coins. He gets 6 more coins. Now he has 13 coins. How many coins did he have to start?
Equation:

_____ coins

3. Gus has some cards. He buys 8 cards. Now he has 15 cards. How many cards did Gus have to start?
Equation:

_____ cards

4. Emma has 5 apples. How many more apples does Emma need to have 17 apples?
Equation:

_____ apples

Copyright © McGraw-Hill Education

5. **Error Analysis** 8 markers are in the bin. More markers go in the bin. Now 12 markers are in the bin. How many more markers went into the bin?

Monica writes 8 + 12 = ? to show the problem. Do you agree? Explain.

6. **Extend Your Thinking** What word problem can you make to match the picture? Make one addend the unknown.

Reflect

How can you show a word problem when one addend is unknown?

Math is... **Mindset**

How did you help identify a problem in your class or community?

Represent and Solve Put Together Problems

Be Curious

What do you see?

Part	Part

Whole

Math is... Mindset

What are your strengths in math?

Learn

A tank holds 4 large fish
and 8 small fish.

How many fish are in the tank?

The **parts** are 4 and 8. The **whole** is unknown. $4 + 8 = ?$	Add the parts. $4 + 8 = 12$

Part	Part
●●●●	●●●● ●●●●
Whole	
?	

Part	Part
●●●●	●●●● ●●●●
Whole	
●●●●●●●● ●●●●	

12 fish are in the tank.

A part-part-whole mat can show
addition problems.

 ## Work Together

7 chicks and 5 ducklings are in a pen.
How many baby birds are in the pen?

Show your thinking.

_____ birds

On My Own

Name _____

1. Ana has 6 tomato plants and 9 cucumber plants.
   How many plants does Ana have?
   Which equation matches the word problem?
   Circle the equation.

   $6 + 9 = ?$ $6 + ? = 9$

What equation shows the problem? Use ? for the unknown. Then solve.

2. There are 8 small paintbrushes
   and 3 large paintbrushes.
   How many paintbrushes?
   Equation:

   _____ paintbrushes

3. Franco has 8 apples. Zoe has 6 apples.
   How many apples do they have?
   Equation:

   _____ apples

4. There are 5 red cars and 7 white cars.
   How many cars in all?
   Equation:

   _____ cars

5. **Error Analysis** Juan and Lucia solve this problem.

> A jar has 9 blue marbles and 6 red marbles.
> How many marbles are in the jar?

Juan writes 6 + 9 = 15. Lucia writes 9 + 6 = 15.
Can they both be correct? Explain.

6. **Extend Your Thinking** Make a word problem to match the part-part-whole mat.

Part	Part
7	7
Whole	
?	

 **Reflect**

How can a part-part-whole mat help you add?

Math is... **Mindset**
How did you use your strengths in math today?

Represent and Solve More Put Together Problems

Be Curious

How are they the same?
How are they different?

Math is... Mindset

What helps you understand a problem situation?

Learn

15 animals are on the beach.
Some are turtles and some are crabs.

How many turtles and how many crabs?

▶ **One Way**

15 = ? + ?

Part	Part
?	?
Whole	
⚫⚫⚫⚫⚫⚫⚫ ⚫⚫⚫⚫⚫⚫⚫⚫	

15 = 7 + 8

There are 7 turtles and 8 crabs.

▶ **Another Way**

15 = ? + ?

Part	Part
?	?
Whole	
⚫⚫⚫⚫⚫⚫⚫⚫⚫⚫ ⚫⚫⚫⚫⚫	

15 = 10 + 5

There are 10 turtles and 5 crabs.

Math is... Quantities

What other equations match the problem?

💬 Work Together

Jia has 19 coins. Some are pennies. The rest are dimes.
How many are pennies and dimes?

_____ pennies

_____ dimes

On My Own

Name _____

What equation shows the problem? Use ? for the unknowns. Then solve.

1. A farm has 11 animals. Some are chickens and some are sheep. How many chickens and how many sheep?
    Equation:

    _____ chickens   _____ sheep

2. There are 6 blue kites and some purple kites. There are 14 kites in all. How many purple kites?
    Equation:

    _____ purple kites

3. There are 16 backpacks. Some are red. The rest are green. How many red backpacks and how many green backpacks?
    Equation:

    _____ red backpacks   _____ green backpacks

4. Gina does 15 exercises. She does some push-ups and 8 sit-ups. How many push-ups does Gina do?
    Equation:

    _____ push-ups

5. Joy has 16 toys. 7 are dolls. The rest are trains. How many are trains?

Draw to show your thinking.

_____ trains

6. Extend Your Thinking Make an equation to match the part-part-whole mat. Use ? for the unknowns. Solve two different ways.

Part	Part
?	?
Whole	
●●●●●●	
●●●●●●●●	

Reflect

What pattern do you notice when you find two unknown addends when you know the total?

Math is... Mindset

What helped you understand a problem situation?

Problems and Equations 1

Name _____

1. There are apples and oranges in a bowl. There are 6 apples and 8 oranges. How many pieces of fruit are in the bowl?

   Circle the equation that shows the problem.

   $6 +$ _____ $= 8$

   $6 + 8 =$ _____

   $8 - 6 =$ _____

   Tell or show why.

2. Lucas has 7 markers. His friend gave him some more markers. Now Lucas has 12 markers. How many markers did his friend give Lucas?

   Circle the equation that shows the problem.

   $7 +$ _____ $= 12$

   $7 + 12 =$ _____

   $12 - 7 =$ _____

   Tell or show why.

3. There are 9 children and 6 adults swimming in a pool. How many people are swimming in the pool?

Circle the equation that shows the problem.

9 + _____ = 6

9 + 6 = _____

9 − 6 = _____

Tell or show why.

Reflect On Your Learning

Represent and Solve Addition Problems with Three Addends

Be Curious

Some bees are on a flower.

Some bees are on another flower.

Some bees are on another flower.

Math is... Mindset

How can you understand thinking that is different from yours?

Learn

4 bees are on a flower.

2 bees are on another flower.

5 bees are on another flower.

How many bees are there?

First, add two addends.

4 + 2 = ? 　　 **+ 2**

4 + 2 = 6

Then add the other addend.

6 + 5 = ?

　　　　　　 + 5

6 + 5 = 11

Math is... Explaining

How is adding three addends like adding two addends?

Some problems have three addends. You add two addends at a time.

Work Together

6 pens are on a desk. 3 pens are in a case.
4 pens are on the floor. How many pens are there?
Show your thinking.

_____ pens

On My Own

Name _____

What two equations show the problem? Use ? for the unknown. Then solve.

1. Cora as 5 cubes. Lin has 4 cubes.
   Brianne has 3 cubes. How many cubes altogether?

   Equation: _____ _____ cubes

   Equation: _____ _____ cubes

2. Kenny has 4 marbles. He finds 5 marbles.
   He buys 9 marbles. How many marbles does
   Kenny have?

   Equation: _____ _____ marbles

   Equation: _____ _____ marbles

3. There are 6 birds in a tree. 3 birds are on a fence.
   7 birds are in the sky. How many birds?

   Equation: _____ _____ birds

   Equation: _____ _____ birds

4. Ami sees 4 cats. She sees 2 dogs.
   She sees 7 fish. How many pets does Ami see?

   Equation: _____ _____ pets

   Equation: _____ _____ pets

5. What equation matches the number line?
Write an equation.

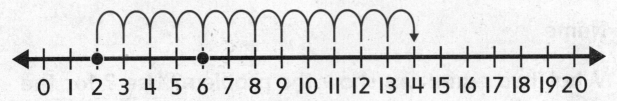

6. Extend Your Thinking What word problem can you make to match the cubes? Write the problem. Then solve.

Reflect

How can you add three addends?

Math is... Mindset

How did understanding thinking that is different from yours help you learn?

Solve Addition Problems

? Be Curious

Some tigers are in the forest.

Some tigers enter the forest.

Now there are more tigers.

Math is... Mindset

What are some ways to resolve disagreements with your classmates?

Learn

Some tigers are in the forest.
5 tigers enter the forest.
Now there are 14 tigers.

How many tigers were in the forest before?

An addition equation shows the problem.

Find the missing addend.

$? + 5 = 14$

Part	Part
?	●●●●●
Whole	
●●●●●●●●●● ●●●●	

$9 + 5 = 14$

Part	Part
●●●●● ●●●●	●●●●●
Whole	
●●●●●●●●●● ●●●●	

9 tigers were in the forest.

You can find different unknowns.

Math is... Choosing Tools

What other tool can you use to find the unknown?

💬 Work Together

There are 13 animals. 7 are rabbits and the rest are birds. How many birds? Show your thinking.

_____ birds

On My Own

Name _____

How can you make an equation to show the problem? Use ? for the unknown. Then solve.

1. A box has 14 blocks.
   7 blocks are large.
   The rest are small.
   How many small blocks?
   Equation:

   _____ small blocks

2. There are 11 boats in the water. 5 more boats
   go in the water. How many boats?
   Equation:

   _____ boats

3. Leo has some books. He gets 7 more books.
   Now he has 15 books. How many books did
   Leo have to start?
   Equation:

   _____ books

4. There are 12 blue pencils. There are 5 red pencils.
   There is 1 green pencil. How many pencils?
   Equation:

   _____ pencils

5. **STEM Connection** Jordan visits a school for guide dogs. There are 14 dogs. More dogs join. Now there are 20 dogs. How many more dogs join?

_____ guide dogs

6. Lola has 12 balloons. Some are big and some are small. How many big balloons and how many small balloons?

Make three different equations to match the problem.

7. **Extend Your Thinking** Make an addition word problem. Then solve.

🔄 **Reflect**

How do you think like a mathematician to solve addition word problems?

Math is... **Mindset**

How have you resolved disagreements with your classmates?

Unit Review

Name _____

Vocabulary Review

Circle the example or examples of the term.

1. part

   $6 + 7 = 13$ $17 = 9 + 8$

2. sum

   $5 - 2 = 3$ $7 + 4 = 11$

3. unknown

   $9 + ? = 12$ $8 + 5 = 13$

4. whole

   $5 + 9 = 14$ $16 = 8 + 8$

5. equation

   $7 + 2 = 9$ $12 = 5 + 7$

Review

6. Owen has 8 toy cars. He gets some more toy cars.
Now he has 12 toy cars. How many toy cars does
Owen get?
Which equation shows the problem?

A. $8 + 12 = ?$

B. $8 + ? = 12$

C. $12 + 8 = ?$

7. Which equation matches the problem? Draw lines.

Jeff chops some carrots.
Lila chops 6 carrots. Together
they chop 11 carrots. How many
carrots does Jeff chop?

$5 + 8 = ?$

Levi peels 5 potatoes. Iris peels
8 potatoes. How many potatoes
do Levi and Iris peel?

$? + ? = 14$

14 pieces of fruit are in a basket.
Some are apples and some are
oranges. How many apples and
how many oranges?

$? + 6 = 11$

8. Some mugs are on a shelf. Joan puts 7 more mugs on the shelf. Now there are 16 mugs on the shelf. How many mugs were on the shelf to start?

_____ mugs

9. Michaela has 12 daisies. She gets 6 tulips. How many flowers does she have?

_____ flowers

10. There are 5 chickens in the coop. 9 chickens are in the yard. 2 chickens are on the fence. How many chickens?

_____ chickens

11. There are 14 swimmers. 9 are boys and the rest are girls. How many swimmers are girls?

_____ girls

Performance Task

A carpenter makes 13 objects to sell.
Some are birdhouses. Some are shelves.

Part A: The carpenter makes 7 birdhouses and some shelves. How many shelves does the carpenter make? Show your thinking.

 shelves

Part B: What other numbers of birdhouses and shelves could the carpenter make? The sum is still 13. Make two equations to show these combinations.

Reflect

How can you solve different addition word problems?

Fluency Practice

Name _____

Fluency Strategy

You can use a ten-frame to help subtract from 10.

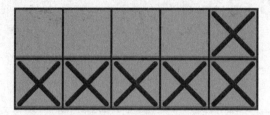

$$10 - 6 = 4$$

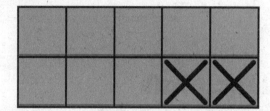

$$10 - 2 = 8$$

1. How can you use the ten-frame to subtract $10 - 7$?
   Show your work.

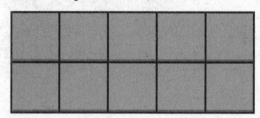

$$10 - 7 = \underline{\quad}$$

Fluency Flash

What is the difference? Use the ten-frame to help.

2.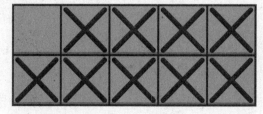

$$10 - 9 = \underline{\quad}$$

3.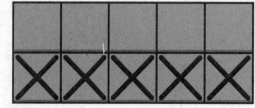

$$10 - 5 = \underline{\quad}$$

Fluency Check

What is the sum or difference? Write the number.

4. $6 + 4 =$ _____

5. $8 - 7 =$ _____

6. $10 - 3 =$ _____

7. $10 - 4 =$ _____

8. $9 - 3 =$ _____

9. $10 - 8 =$ _____

10. $3 + 7 =$ _____

11. $7 - 5 =$ _____

12. $9 + 1 =$ _____

13. $2 + 8 =$ _____

Fluency Talk

How can you show that $10 - 1 = 9$?
Explain your work.

How can you use a ten-frame to add $3 + 7$?
Explain your thinking.

Meanings of Subtraction

Focus Question

How can I solve subtraction problems?

Hi! Ruby here,

I want to be a veterinarian. I have a bag with 14 puppy treats. After I feed the puppies, I have 5 treats left. I can subtract to find out how many treats I gave them.

Name

Make Fifteen

Use Number Cards 0–9. Stack the number cards face up. Take turns choosing cards. Keep them face up. To win, make a sum of 15 with 3 cards.

You can use unit cubes with these five-frames or the number line to help you.

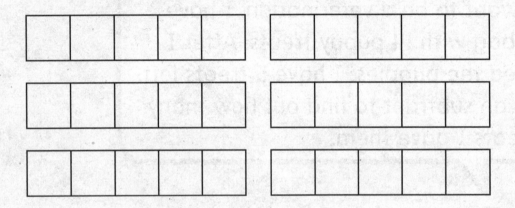

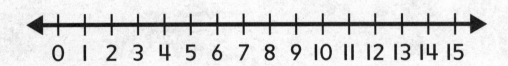

0	1	2	3	4
5	6	7	8	9

Represent and Solve Take From Problems

Be Curious

What question could you ask?

Math is... Mindset

What helps you want to do your best work?

Learn

Juanita has 12 sheep.
She gives 4 sheep away.

**How many sheep does
Juanita have now?**

You can use counters to show the sheep.	An equation shows the problem.
12 counters	$12 - 4 = ?$

Cross out 4 counters.

Math is... Choosing Tools

What is another tool you can use?

$12 - 4 = 8$
Juanita has 8 sheep.

When you subtract, you take away.

💬 Work Together

There are 15 balloons. 8 balloons pop.
How many balloons are left?

Show your thinking.

_____ balloons

On My Own

Name _____

1. There are 9 cats. 5 cats leave.
    How many cats are there now?
    Which equation matches the word problem?
    Circle the equation.

$9 + 5 = ?$ $9 - 5 = ?$

Write an equation to show the problem. Use ? for the unknown. Then solve.

2. Lee has 14 puzzles. He gives away 7 puzzles.
    How many puzzles does Lee have now?

    Equation: _____

    _____ puzzles

3. There are 15 horses in a stable. 6 horses go outside.
    How many horses stay?

    Equation: _____

    _____ horses

4. **Error Analysis** 11 oranges are in the basket. 7 oranges get eaten. How many oranges are left?

Leah makes the equation ? − 7 = 11 to show the problem. Do you agree? Explain.

5. **Extend Your Thinking** Make up a word problem where the difference is the unknown. Draw to solve the problem.

⟳ Reflect

How do you know when to subtract to solve a problem?

Math is... **Mindset**
What helped you want to do your best work?

Represent and Solve More Take From Problems

?

Be Curious

Some apples are on a tree.
A horse eats some apples.
Some apples are left on the tree.

Math is... **Mindset**

What helps you understand how others are feeling?

Learn

16 apples are on a tree. A horse eats some apples. 9 apples are left on the tree.

How many apples did the horse eat?

The cubes can show the apples.  16 cubes	An equation shows the problem. $16 - ? = 9$ 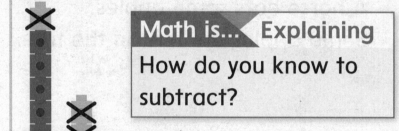 Cross out cubes until 9 are left. $16 - 7 = 9$ The horse eats 7 apples.

Math is... Explaining

How do you know to subtract?

Sometimes when you subtract, you find how many are taken away.

🗨 Work Together

14 ducks are in a pond. Some fly away. Now 11 ducks are in the pond. How many ducks fly away?

Show your thinking.

_____ ducks

On My Own

Name_____

Write an equation to show the problem. Use ? for the unknown. Then solve.

1. A clown has 12 balloons.
   Some balloons pop.
   Now there are 8 balloons.
   How many balloons popped?

   Equation: _____

   _____ balloons

2. There are 17 students on the bus. Some students get off the bus. Now 9 students are on the bus. How many students get off the bus?

   Equation: _____

   _____ students

3. There are 16 grapes in a bowl. Kevin eats some grapes. Now there are 10 grapes in the bowl. How many grapes did Kevin eat?

   Equation: _____

   _____ grapes

4. STEM Connection Ruby sees 15 dogs in the kennel. Then some dogs go home. Now 7 dogs are in the kennel. How many dogs went home?

_____ dogs

5. Extend Your Thinking Make a word problem with an unknown change number. Then solve.

⟳ Reflect

How do you know where to put the unknown in a subtraction equation?

Math is... **Mindset**

What has helped you understand how others are feeling?

Be Curious

What do you notice?
What do you wonder?

Math is... **Mindset**
How well do you think you will do with today's tasks?

Learn

Henry has some shirts in a basket.
He takes out 5 shirts. 7 shirts remain.

How many shirts did Henry have in the basket?

The parts are 5 and 7.
The whole is unknown.

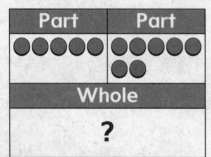

Part	Part
●●●●●	●●●●●
	●●
Whole	
?	

Math is... Thinking

How can you use addition to show the problem?

An equation shows the problem.

$? - 5 = 7$

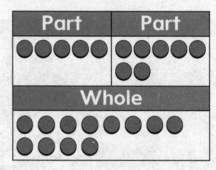

Part	Part
●●●●●	●●●●●
	●●
Whole	
●●●●●●●	
●●●●●	

$12 - 5 = 7$

Henry had 12 shirts in the basket.

In some subtraction equations, the whole is unknown.

🗨 Work Together

Some purple and green grapes are in a bowl.
8 grapes are purple. 11 grapes are green.
How many grapes are in the bowl?

Show your thinking.

_____ grapes

On My Own

Name _____

Write an equation to show the problem. Use ? for the unknown. Then solve.

1. Rai has some blocks.
   He takes out 4 blocks, 10 blocks are left.
   How many blocks did Rai have?

   Equation: _____

   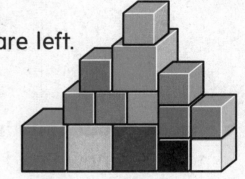

   _____ blocks

2. A bookshelf holds some books.
   8 books are on the bottom shelf.
   9 books are on the top shelf.
   How many books?

   Equation: _____

   _____ books

3. Some caterpillars are in a tree.
   8 are green and 5 are yellow.
   How many caterpillars?

   Equation: _____

   _____ caterpillars

4. Simon has some ribbons in a bag.
He takes 5 ribbons out. 7 ribbons stay in the bag.
How many ribbons does Simon have?
Draw to show your thinking.

_____ ribbons

5. **Extend Your Thinking** Make a problem. Make the total unknown and the two parts known.
Then solve.

🔄 Reflect

How can you show a problem when the whole is unknown?

Math is... Mindset
How well do you think you did with today's tasks?

Problems and Equations 2

Name _____

1. Alex has 11 dimes. He gives 7 dimes to his sister. How many dimes does Alex have left?

   Circle the equation that shows the problem.

   11 − ____ = 7

   11 − 7 = ____

   11 + 7 = ____

   Tell or show why.

---

2. Mia has 13 stars on her paper. Then she glues on some more stars. Now she has 19 stars on her paper. How many stars did she just glue on?

   Circle the equation that shows the problem.

   19 − ____ = 13

   19 − 13 = ____

   13 + ____ = 19

   Tell or show why.

3. 17 cars are in a lot.
Some cars drive away. Now
8 cars are in the lot. How
many cars drive away?

**Circle the equation that
shows the problem.**

17 − ____ = 8

17 − 8 = ____

17 + 8 = ____

Tell or show why.

<hr>

Reflect On Your Learning

Represent and Solve More Take Apart Problems

Be Curious

Tell me everything you can.

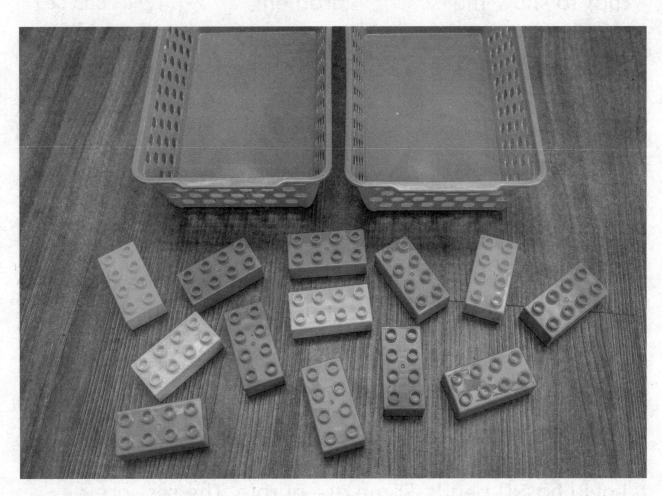

Math is... Mindset

What helps you know when there is a problem?

Learn

Jill has 13 balls in a bin. She puts some in a drawer. The rest stay in a bin.

How many can be in the drawer and how many can be in the bin?

Part	Part
?	?
Whole	

Use a part-part-whole mat to show the problem.

> **Math is...** Quantities
>
> What other equations match the problem?

An equation can show the problem.

$13 - ? = ?$

> What can be one part?

Part	Part
●●● ●●●	●●●● ●●●
Whole	

$13 - 6 = 7$

6 balls can be in the drawer and 7 can be in the bin.

When both parts of an equation are unknown, there can be many different equations.

Work Together

Layla has 10 beads. Some are purple. The rest are orange. How many purple and how many orange?

Show your thinking.

_____ purple beads and _____ orange beads

On My Own

Name _____

Write an equation to show the problem. Use ? for the unknown. Then solve.

1. Carlita has 11 hair ties. Some are old and some are new. How many are old and how many are new?

   Equation: _____

   _____ old and _____ new

2. There are 15 balls. Some roll left. The others roll right. How many roll left and how many roll right?

   Equation: _____

   _____ roll left and _____ roll right

3. There are 19 bugs. Some are ladybugs. Some are beetles. How many ladybugs and how many beetles?

   Write two different answers.

   _____ ladybugs and _____ beetles

   _____ ladybugs and _____ beetles

4. Error Analysis Micah and Lena solve this problem.

> There are 15 bracelets and necklaces on a tray. How many bracelets and how many necklaces?

Micah writes $15 - 10 = 5$. Lena writes $15 - 8 = 7$. Can they both be correct? Explain.

5. Extend Your Thinking Make a subtraction problem to match the part-part-whole mat.

Part	Part
?	?
Whole	
13	

Reflect

How do the two unknown parts in a problem relate to the total?

Math is... Mindset
What helped you know when there was a problem?

Be Curious

Raj has some mugs.
He puts some mugs in the cabinet.
He puts the rest on the table.

Math is... Mindset
What makes you feel frustrated in math?

Learn

Raj has 14 mugs on the table.
He puts 9 mugs in the cabinet.

How many mugs are still on the table?

▶ **One Way** Use a part-part-whole mat.

14 − 9 = ?

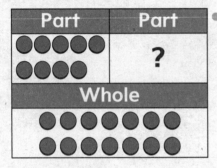

Think addition
to subtract.

Math is... **Perseverance**

How can you check that
your answer is correct?

▶ **Another Way** Use a number line.

14 − 9 = ?

14 − 9 = 5

5 mugs are still on
the table.

Count back
to subtract.

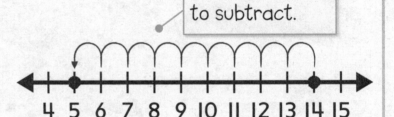

4 5 6 7 8 9 10 11 12 13 14 15

You can show a problem in different ways.

Work Together

Hannah has 15 dishes. 6 are plates and the rest are
bowls. How many bowls does she have?

Show your thinking.

_____ bowls

On My Own

Name _____

Write an equation to show the problem. Use ? for the unknown. Then solve.

1. There are 12 pieces of fruit. 4 are bananas and the rest are apples. How many apples?

Equation: _____

_____ apples

2. Jian has 13 fish. 6 are small and the rest are large. How many large fish does Jian have?

Equation: _____

_____ large fish

3. Stew sees 19 turtles. He sees 7 box turtles. The rest are sea turtles. How many sea turtles does Stew see?

Equation: _____

_____ sea turtles

4. **STEM Connection** Chloe builds
11 birdhouses. She paints 6 blue.
She paints the rest yellow. How
many does she paint yellow?
Draw to show your thinking.

_____ birdhouses

5. **Extend Your Thinking** Make up
a subtraction problem to match
the part-part-whole mat.

Part	Part
●●●● ●●●●	**?**
Whole	
●●●●●●●● ●●●●●●●	

🎨 Reflect

How did you think like a mathematician to solve
subtraction word problems?

Math is... **Mindset**
What made you feel
frustrated in math today?

Solve More Problems Involving Subtraction

? Be Curious

The coach has some basketballs.
She puts some in a bin.
She has some basketballs left.

> **Math is... Mindset**
> How can different ideas help you learn better?

Learn

The coach has 16 basketballs.
She puts some in a bin.
She has 8 basketballs left.

How many basketballs does the coach put in the bin?

Use a part-part-whole mat.

$16 - ? = 8$

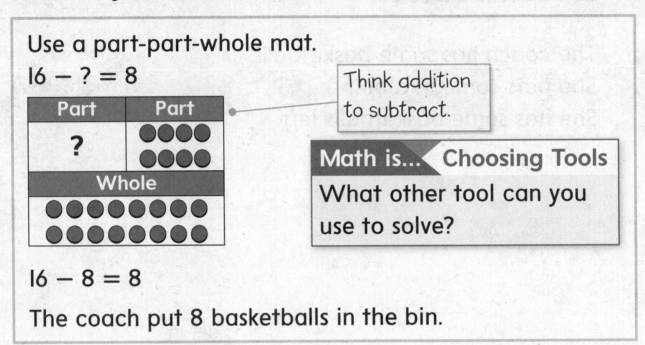

Think addition to subtract.

Math is... Choosing Tools

What other tool can you use to solve?

$16 - 8 = 8$

The coach put 8 basketballs in the bin.

Either part in a subtraction equation can be unknown.
You can find different unknowns.

Work Together

There are 14 animals. Some are zebras and 3 are lions. How many zebras?

Show your thinking.

_____ zebras

On My Own

Name _____

Write an equation to show the problem. Use ? for the unknown. Then solve.

1. Kelly has 16 carrots.
   She cooks 8 carrots for dinner.
   How many carrots are left?

   Equation: _____

   _____ carrots

2. Ann has some apples. She eats 7 apples.
   She has 10 apples left. How many did she
   have to start?

   Equation: _____

   _____ apples

3. There are 13 squirrels. Some run away. 4 squirrels
   are left. How many squirrels ran away?

   Equation: _____

   _____ squirrels

4. **STEM Connection** C.J. asks 14 students to name their favorite sport. 9 say baseball and the rest say football. How many students said football?

Draw to show your thinking.

____ students

5. **Extend Your Thinking** Make a subtraction problem where both parts are known but the total is unknown.

⊙ Reflect

How can you show different kinds of subtraction problems?

Math is... **Mindset**

How have different ideas helped you learn better?

Solve Problems Involving Addition and Subtraction

? Be Curious

What could the question be?

Math is... Mindset

What can you do today to help build a good relationship with a classmate?

Learn

There are 14 cherries. Emma takes some cherries and leaves her brother 8 cherries.

How many cherries does Emma take?

▶ **One Way** Use a subtraction equation.

$14 - ? = 8$

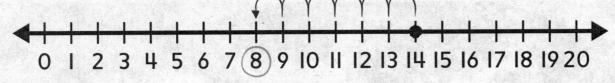

$14 - 6 = 8$

▶ **Another Way** Use an addition equation.

$14 = ? + 8$

$14 = 6 + 8$ Emma takes 6 cherries.

You can add or subtract to solve some problems.

Math is... Explaining

Which operation will you use?

🗨 Work Together

Wes has 11 grapes. He eats some grapes. Now he has 4 grapes. How many grapes did Wes eat?
Show your thinking.

_____ grapes

On My Own

Name _____

Write an equation to show the problem. Use ? for the unknown. Then solve.

1. Kendra makes 15 hotdogs.
8 hotdogs are on a plate.
The rest are on the grill. How
many hotdogs are on the grill?

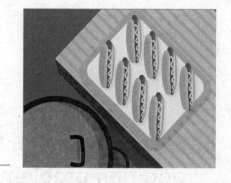

Equation: _____

_____ hotdogs

2. There are 11 hippos. More hippos join. Now
there are 17 hippos. How many hippos join?

Equation: _____

_____ hippos

3. There are 18 magnets. 9 magnets are squares and
the rest are circles. How many magnets are circles?

Equation: _____

_____ magnets

4. Error Analysis Abe and Leo solve this problem.

> There are 13 cartons of milk. Students drink 6 of the cartons of milk. How many cartons of milk are left?

Abe writes $13 - 6 = ?$ and Leo writes $6 + ? = 13$.

Can they both be correct? Explain.

5. Extend Your Thinking Make a subtraction or addition problem. Then solve. Draw to show your work.

 **Reflect**

How can you add or subtract to solve problems?

Math is... Mindset

What have you done today to help build a good relationship with a classmate?

Unit Review

Vocabulary Review

Use the vocabulary to complete the sentence.

difference	part
unknown	whole
problem	

1. A ? shows the _____ in an equation.

2. You subtract one number from another to find the _____ .

3. In 12 – 7 = 5, 7 is one _____ and 5 is the other.

4. In a subtraction problem, the total is the same as the _____ .

5. A _____ tells a story that uses numbers.

Review

6. There are 13 stuffed animals. Some are tigers. 8 are zebras. How many are tigers?

   Which equation shows the problem? Choose all the correct answers.

   A. $13 + 8 = ?$

   B. $8 + ? = 13$

   C. $13 - ? = 8$

7. Bonnie reads 12 books. Some are animal books and some are science books. How many of each book does Bonnie read?

   Which equation shows the problem? Choose all the correct answers.

   A. $12 = 5 + 7$

   B. $7 + 8 = 15$

   C. $9 + 3 = 12$

8. There are 12 dogs at the park. Some dogs go home. Now 3 dogs are at the park. How many dogs went home?

   _____ dogs

9. Jake has 10 books in his backpack. He takes out 3 books. How many books are left?

_____ books

10. A jar holds orange and blue marbles. There are 5 orange marbles and 9 blue marbles. How many marbles are in the jar?

_____ marbles

11. Hannah has 12 tools. 4 are hammers and the rest are wrenches. How many wrenches does she have?

_____ wrenches

12. There are 15 kites. Some are flat and some are boxy. How many of each kite could there be? Choose all the correct answers.

    A. 5 flat and 7 boxy

    B. 6 flat and 9 boxy

    C. 7 flat and 8 boxy

13. Evan has some boxes to move. He moves 6 boxes. He has 5 more boxes to move. How many boxes did Evan start with?

_____ boxes

Performance Task

The veterinarian sees fewer than 20 animals today. He sees 10 dogs before lunch.

Part A: Does the veterinarian see the same number of dogs and cats today? Explain your thinking.

Part B: How many cats can the veterinarian see?

Part C: What can you say about the number of dogs compared to the number of cats the veterinarian sees?

 **Reflect**

How can you solve different subtraction problems?

Fluency Practice

Name _____

Fluency Strategy

> You can add 0 to a number. The sum is that number. You can subtract 0 from a number. The difference is that number.
>
>  $+$ [ ]
>
> $3 + 0 = 3$
>
>  $-$ [ ]
>
> $2 - 0 = 2$

1. How can you find the sum of $8 + 0$? How can you find the difference of $8 - 0$? Explain your thinking.

Fluency Flash

What is the sum or difference? Write the number.

2. $4 + 0 =$ _____

3. $1 - 0 =$ _____

Fluency Check

What is the sum or difference? Write the number.

4. $10 - 3 = $ _____

5. $8 + 2 = $ _____

6. $9 - 0 = $ _____

7. $7 + 0 = $ _____

8. $5 + 5 = $ _____

9. $3 + 0 = $ _____

10. $10 - 0 = $ _____

11. $1 + 9 = $ _____

12. $4 - 0 = $ _____

13. $6 + 0 = $ _____

Fluency Talk

What do you know about $5 + 0$? How can you use what you know to add 0 to other numbers?

How can you show $10 - 6 = 4$? Explain your work.

Addition within 100

Focus Question

How do I use strategies to add 2-digit numbers?

Hi, I'm Marisol.

I want to be a paramedic. Paramedics ride in ambulances to help people. I bet they ride a lot of miles. How can they find out how many miles they ride each week?

STEM video | GO ONLINE

Race to 21

Play an adding game.

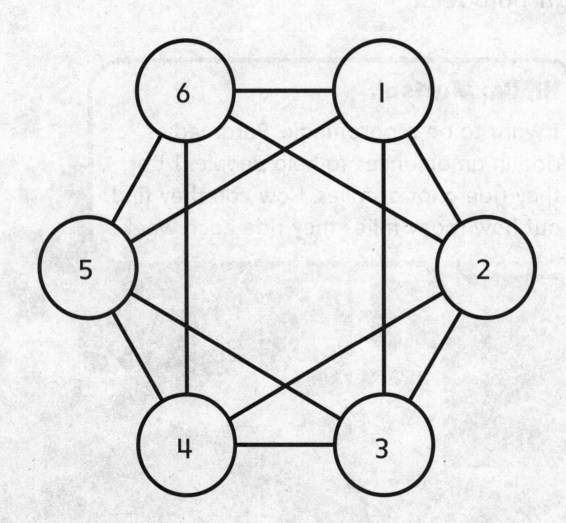

Use Mental Math to Find 10 More

Be Curious

What do you see?

Math is... Mindset

How can you be part of the classroom community?

Learn

Rachel has 21 stamps. Henry gives her 10 more stamps.

How many stamps does Rachel have now?

You can use mental math to add 10 to a **2-digit number**.

$$21 + 10 = 31$$

$$47 + 10 = 57$$

$$74 + 10 = 84$$

Math is... Patterns

What patterns do you see?

When you add 10, the tens **digit** goes up by 1 and the ones digit stays the same.

Work Together

Tom has 35 grapes. He gets 10 more grapes. How many grapes does Tom have now?

Explain your thinking.

_____ grapes

On My Own

MATH REPLAY | GO ONLINE

Name ______________________________

Is the equation true? Circle Yes or No.

1. 17 + 10 = 18

   Yes No

2. 32 + 10 = 42

   Yes No

What is the sum?

3. 10 + 51 = ____

4. 86 + 10 = ____

5. 10 + 64 = ____

6. 79 + 10 = ____

7. 18 + 10 = ____

8. 44 + 10 = ____

9. Brenda has 58 red ribbons and 10 blue ribbons. How many ribbons does she have?

_____ ribbons

10. **STEM Connection** Ruby gives her dog 16 treats each day. The vet says the dog can have 10 more treats each day. How many treats can Ruby's dog have each day?

_____ treats

11. **Extend Your Thinking** Some crayons are on the table. 10 crayons are in a box. There are 33 crayons in all. How many crayons are on the table? Explain your thinking.

_____ crayons

 **Reflect**

How can you add 10 to a number?

Math is... **Mindset**

How were you part of the classroom community today?

Patterns and Addition

CHERYL TOBEY
MATH
PROBES

Name _____

Circle the answer.

I. What number belongs in the ? box in the chart?

	57	58	
		?	
76			

59   62   68   72   74   75

None of these numbers

Tell or show why.

---

2. What number belongs in the ? box in the chart?

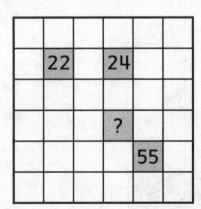

25   26   36   44   50   54

None of these numbers

Tell or show why.

3. What is 10 more than 74?

Tell or show why.

75   76   80   84   85

None of these numbers

4. What is 20 more than 43?

Tell or show why.

45   53   60   63   65

None of these numbers

Reflect On Your Learning

Represent Adding Tens

? Be Curious

Tell me everything you can.

Math is... Mindset

What helps you make good decisions about your behavior?

Learn

Milo has 48 trading cards.
He buys 30 more trading cards.

How many trading cards does Milo have now?

48 + 30 = ?

Show 48 and 30.

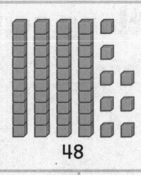

48   30

Add the tens.

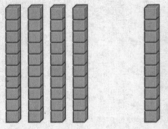 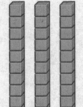

40   +   30   = 70

Math is... **Patterns**

How is adding 30 similar to adding 10?

Add the tens and ones.

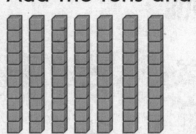

70   +   8 = 78

So, 48 + 30 = **78**.

Work Together

What is the sum?
Show your thinking.

20 + 63 = _____

On My Own

Name _____

What is the sum? Fill in the equation.

1.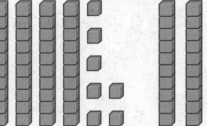

    ___ + ___ = ___

2.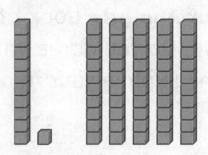

    ___ + ___ = ___

What is the sum?

3. $29 + 60 =$ ____

4. $35 + 40 =$ ____

5. $78 + 10 =$ ____

6. $30 + 69 =$ ____

7. Pete sees 25 fish. He sees 20 more fish.
   How many fish does Pete see in all?

   _____ fish

8. **Error Analysis** Look at Min's
   work. Do you agree with the
   sum? Explain your thinking.

   $$19 + 40 = 50$$

9. **Extend Your Thinking** Write a word problem about
   adding tens to a number. Then solve.

⟳ Reflect

How can you add tens to any number?

Math is... Mindset
What helped you
make good decisions
about your behavior?

Represent Adding Tens and Ones

Be Curious

What do you notice?
What do you wonder?

Math is... Mindset

What are your math superpowers?

Learn

Don has 22 oars. He buys 6 more oars.

How many oars does Don have now?

$22 + 6 = ?$

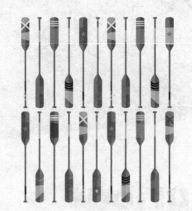

Show 22 and 6.

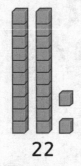

22   6

Math is... Explaining

Why do you add 6 to the ones and not to the tens?

Add the ones.

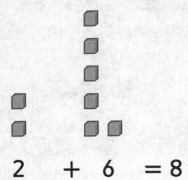

$2 + 6 = 8$

Add the tens and ones.

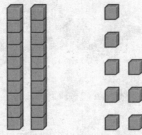

$20 + 8 = 28$

So, $22 + 6 = $ **28**.

Work Together

What is the sum? Show your thinking.

$62 + 5 = $ _____

On My Own

Name _____

What is the sum? Fill in the equation.

1.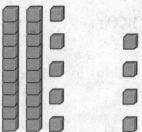

   ____ + ____ = ____

2.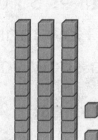

   ____ + ____ = ____

What is the sum?

3. 34 + 4 = _____

4. 65 + 2 = _____

5. 3 + 26 = _____

6. 51 + 8 = _____

7. The farmer has 21 melons. She picks 6 more melons. How many melons does the farmer have now?

_____ melons

8. Laura takes 35 photos. She takes 3 more photos. How many photos does Laura take? Draw to show your thinking.

_____ photos

9. Extend Your Thinking Make your own equation adding ones to a 2-digit number. Then solve.

⊙ **Reflect**

How can you add ones to a number?

Math is... Mindset
How have your math superpowers helped you today?

Decompose Addends to Add

? Be Curious

What do you notice?
What do you wonder?

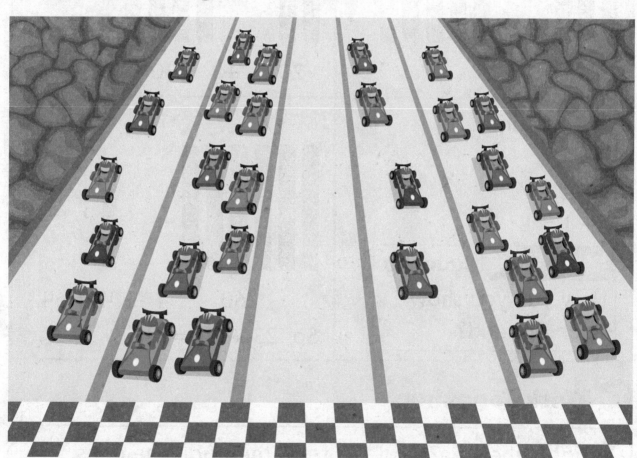

Copyright © McGraw-Hill Education

Math is... Mindset

How can you show you understand how others are feeling?

Learn

23 cars are blue. 41 cars are red.

How many cars are there?

23 + 41 = ?

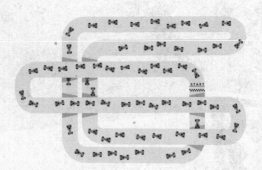

Show 23 and 41.	Add the tens.
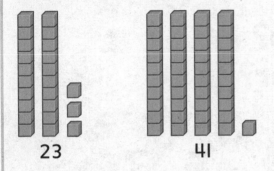 23 41	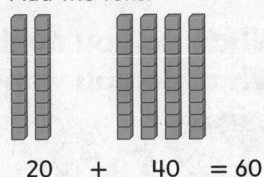 20 + 40 = 60

Add the ones.

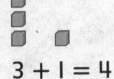

3 + 1 = 4

Math is... Modeling
How can you show each addend?

60   + 4 = 64

So, 23 + 41 = **64**.

Work Together

What is the sum? Add the tens, then add the ones.
Then add the tens and ones.

34 + 25 = _____

On My Own

Name _____

What is the sum?

1.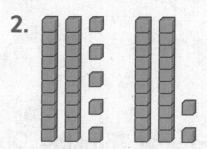

   33 + 43 = _____

2. 25 + 22 = _____

What is the sum? Show adding the tens, then the ones. Then add the tens and ones.

3. 61 + 18 = _____

4. 43 + 25 = _____

5. 21 + 51 = _____

6. 14 + 34 = _____

7. **Error Analysis** Val adds 31 + 27.

$$1 + 7 = 8$$

$$30 + 20 = 50$$

She says the sum is 50. Do you agree? Explain.

8. **Extend Your Thinking** One team has 26 players. One team has 23 players. How many players are on the two teams?

 players

Reflect

How is adding tens then ones helpful?

Math is... Mindset

What helped you understand how others are feeling?

Use an Open Number Line to Add within 100

? Be Curious

How are they the same?
How are they different?

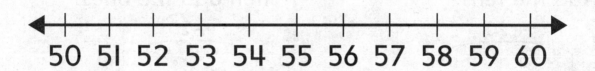

50 51 52 53 54 55 56 57 58 59 60

55

Math is... Mindset

What helps you solve a problem?

Learn

How can you use a number line to add 55 + 24?

55 + 24 = ?

Start with 55. It is the greater addend. 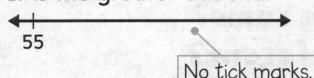 No tick marks.	**Math is...** **Explaining** Why do you start with the greater addend?

| Add the tens.
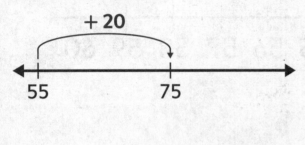 | Then add the ones.
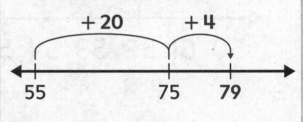
55 + 24 = **79** |

You can use an **open number line** to add numbers.

Work Together

What is the sum? Use an open number line.

13 + 26 = _____

On My Own

Name _____

What is the sum?

1. $24 + 34 =$ _____

2. $72 + 15 =$ _____

3. $37 + 40 =$ _____

4. $63 + 21 =$ _____

What is the sum? Use the open number line.

5. $65 + 21 =$ _____

6. $28 + 51 =$ _____

7. **STEM Connection** Paramedics drive
   35 miles. They drive 22 more miles.
   How many miles do they drive?
   Show your thinking on the
   open number line.

   _____ miles

   <-->

8. **Extend Your Thinking** How is adding on an open
   number line like counting on to add?

Reflect

How can you use an open number line to add?

Math is... Mindset

What helped you solve
a problem?

Decompose to Add on an Open Number Line

Be Curious

What could the question be?

Math is... Mindset

What is your goal for today?

Learn

Jill has 35 pompoms.
Daryn has 8 pompoms.

How many pompoms do they have?

$35 + 8 = ?$

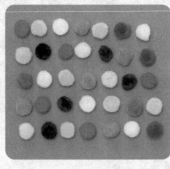
Jill

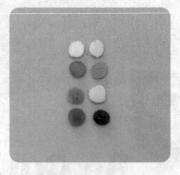
Daryn

▶ **One Way** Count on by 1.

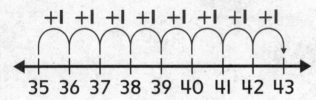

▶ **Another Way** Make a 10 to add.

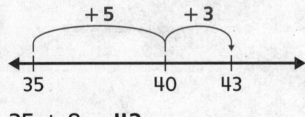

$35 + 8 = 43$

> **Math is...** Quantities
>
> How does the number line show each addend?

When you add, sometimes you break apart one addend to make a 10.

🗨 Work Together

What is the sum? Use the open number line.

$86 + 7 =$ _____

On My Own

Name _____

What is the sum? Use the open number line.

1. $27 + 4 =$ _____

2. $45 + 8 =$ _____

What is the sum?

3. $9 + 62 =$ _____   4. $36 + 7 =$ _____

5. $59 + 5 =$ _____   6. $6 + 74 =$ _____

7. Maria has 37 round beads. She has 9 long beads. How many beads does Maria have?

_____ beads

8. **Error Analysis** Enzo adds 9 + 23.

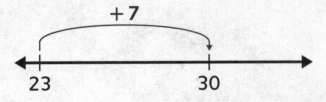

+7

23   30

He says the sum is 30. Do you agree? Explain.

9. **Extend Your Thinking** Kia has 26 buttons. She buys 7 more. How can you find how many buttons?

↻ **Reflect**

How can you use an open number line to add ones to a number?

Math is... **Mindset**
How have you worked to reach your goal today?

Regroup to Add

Be Curious

What do you notice?
What do you wonder?

Math is... Mindset

How do you share
your thinking clearly?

Learn

46 toys are in the machine.
7 toys are near it.

How many toys are there?

46 + 7 = ?

Add the ones.

Make a 10.

46   7

Math is... Explaining

How many tens and ones are in 46 and 7?

Regroup 10 ones as 1 ten.

50   3

Add the tens and ones.

50   +   3 = 53

46 + 7 = **53**

💬 Work Together

What is the sum? Explain your thinking.

34 + 8 = _____

On My Own

MATH REPLAY | GO ONLINE

Name ______________________

What is the sum? Show regrouping 10 ones as 1 ten.

1. 36 + 5 = _____

2. 65 + 8 = _____

What is the sum?

3. 82 + 9 = _____

4. 6 + 79 = _____

5. 58 + 4 = _____

6. 7 + 47 = _____

7. **Error Analysis** Marion uses blocks to add 38 + 8.

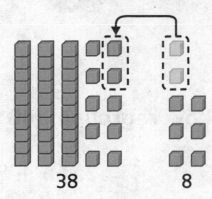

38   8

Marion says the sum is 48. How can you help
Marion add correctly?

8. **Extend Your Thinking** Write an addition problem
where you have to regroup to add. Then explain
how to solve your addition problem.

 **Reflect**

How can you regroup to add ones to a number?

Math is... Mindset

Why is it important to share
your thinking clearly?

Add 2-Digit Numbers

Be Curious

What question could you ask?

Math is... Mindset

What do you want your classmates to know about your math story?

Learn

Macy has 47 pins on her backpack.
She puts 15 pins on the pocket.

**How many pins does Macy have
on her backpack?**

47 + 15 = ? 47　　　　15	Add the tens. 40 + 10 = 50
Add the ones.　Regroup 　　　　　　　10 ones 　 = 　　　　as 1 ten. 7 + 5 = 12	Add the tens and ones. 60　+　2 47 + 15 = 62

When you add, sometimes you
regroup 10 ones as 1 ten. Then
you add the tens and ones.

Math is... Explaining

When do you regroup
ones as tens?

🗨 Work Together

What is the sum?

28 + 54 = _____

On My Own

Name _____

What is the sum? Draw tens and ones to show your thinking.

1. 28 + 12 = _____

2. 14 + 19 = _____

///

What is the sum?

3. 15 + 59 = _____

4. 39 + 23 = _____

5. 48 + 28 = _____

6. 37 + 58 = _____

7. **STEM Connection** Jordan feeds the horses for 25 minutes. He brushes the horses for 49 minutes. How many minutes does Jordan stay with the horses?

_____ minutes

8. **Extend Your Thinking** How do you know when to regroup when you add 2-digit numbers?

Reflect

How can you add 2-digit numbers with regrouping?

Math is... **Mindset**

What did you tell your classmates about your math story?

Unit Review

Name _____

Vocabulary Review

Use the vocabulary to complete each sentence.

2-digit number	digit
ones	open number line
regroup	tens

1. A _____ is a symbol used to write numbers.

2. You can _____ 10 ones as 1 ten.

3. This is an _____.

4. The number 54 has 4 _____.

5. A number with 2 digits is called a _____.

6. The number 38 has 3 _____.

Review

7. What is the sum? Show your thinking on the number line.

$37 + 21 = $ _____

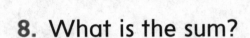

8. What is the sum?

$24 + 10 = $ _____

9. What is the sum?

$15 + 40 = $ _____

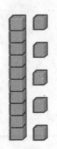 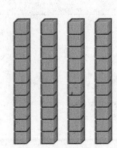

10. What is the sum? Draw to show your thinking.

$21 + 7 = $ _____

11. What is the sum of 21 + 34? Choose the correct answer.

    A. 50

    B. 51

    C. 55

12. What is the sum? Show your thinking on the number line.

    34 + 8 = _____

13. What is the sum? Draw to show your thinking.

    23 + 49 = _____

14. What is the sum? Draw to show regrouping.

    54 + 7 = _____

Performance Task

Paramedics help many people every day.

Part A: They help 22 people Monday and 37 people Tuesday. How many people do they help on the two days?

_____ people

Part B: On Wednesday they help 16 people before lunch. They help 9 people after lunch. How many people do they help on Wednesday?

_____ people

Part C: They help 34 people Thursday. They help 48 people Friday. How many people do they help on the two days?

_____ people

 **Reflect**

What are some ways to add 2-digit numbers?

Fluency Practice

Name _____

Fluency Strategy

You can use doubles to add.

$4 + 4 = ?$

 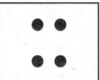

Count the dots. There are 8.

So, $4 + 4 = 8$.

1. How can you use doubles to add $2 + 2$?
   Draw dots and explain.

Fluency Flash

What is the sum? Write the number.

2. $3 + 3 =$ _____

3. $5 + 5 =$ _____

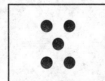

Fluency Check

What is the sum or difference? Write the number.

4. $8 - 0 = $ _____

5. $1 + 1 = $ _____

6. $2 + 2 = $ _____

7. $10 - 2 = $ _____

8. $4 + 4 = $ _____

9. $10 - 4 = $ _____

10. $5 + 0 = $ _____

11. $10 - 7 = $ _____

12. $3 + 3 = $ _____

13. $9 - 0 = $ _____

Fluency Talk

How can you use doubles to add $1 + 1$?

How can you show $5 - 0 = 5$? Explain your work.

Compare Using Addition and Subtraction

Focus Question

How can I compare using addition and subtraction?

Hi, I'm Emily.

I want to be an aerospace engineer. I know one airplane holds 16 people. Another holds only 8. I can subtract to find out how many more people the bigger plane can hold.

STEM video | GO ONLINE

Name _____

Three Numbers in Order

A. What are these sums?

$1 + 3 =$ _____ $2 + 2 =$ _____

B. Pick three numbers that are in order.
List them below.

Numbers: _____ _____ _____

What is the sum
of the least and
greatest numbers?

_____ + _____ = _____

What is the sum of
the middle number
added to itself?

_____ + _____ = _____

C. Pick three numbers that are 2 apart as you go in
order (such as 1, 3, 5). List them below.

Numbers: _____ _____ _____

What is the sum
of the least and
greatest numbers?

_____ + _____ = _____

What is the sum of
the middle number
added to itself?

_____ + _____ = _____

Represent and Solve Compare Problems

Be Curious

Tell me everything you can.

Math is... Mindset

What are some ways you can connect with your classmates?

Learn

Jordan plants 11 flowers.
Chey plants 17 flowers.

How many more flowers does Chey plant than Jordan?

Cubes can show the problem.

Jordan's flowers

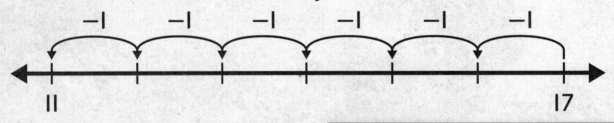

Chey's flowers

more than Jordan

$17 - 11 = ?$

Count back to find how many more.

-1 -1 -1 -1 -1 -1

11   17

$17 - 11 = 6$

Chey has 6 more flowers.

Math is... **Connections**

How can you use addition to solve?

You can subtract to compare numbers.

Work Together

Joy sees 18 bugs. Alex sees 12 bugs. How many fewer bugs does Alex see than Joy? Show your thinking.

_____ bugs

On My Own

Name _____

What equation can show the problem? Use ? for the unknown. Then solve.

I. Roy has 13 markers. Joyce has 7 markers. How many more markers does Roy have than Joyce?

Roy

Joyce

Equation: _____

_____ markers

2. Oscar collects 4 leaves. Sammy collects 8 leaves. How many fewer leaves does Oscar collect?

Equation: _____

_____ leaves

3. Natasha has II books. Sheila has 19 books. How many fewer books does Natasha have than Sheila?

Equation: _____

_____ books

4. **STEM Connection** A plane holds 18 people. A helicopter holds 15 people. How many more people does the plane hold?

Draw to show your thinking.

_____ people

5. **Extend Your Thinking** Make a word problem to compare two numbers. Use the word *fewer*.

Reflect

How can you compare numbers to find the difference?

Math is... Mindset

How did you connect with your classmates today?

Represent and Solve Compare Problems Using Addition

Be Curious

What do you notice?
What do you wonder?

Math is... Mindset

What do you need to be ready to learn?

Learn

Jamal has 4 more crackers than Emma. Emma has 9 crackers.

How many crackers does Jamal have?

Counters can show the problem.

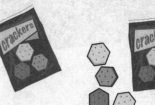
Emma   Jamal

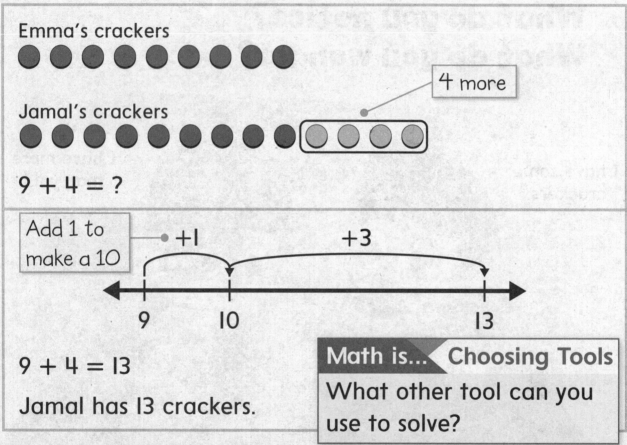

Emma's crackers

4 more

Jamal's crackers

$9 + 4 = ?$

Add 1 to make a 10

+1   +3

9   10   13

$9 + 4 = 13$

Jamal has 13 crackers.

Math is... Choosing Tools

What other tool can you use to solve?

You can add to find how many more.

🗨 Work Together

Pete has 6 fewer grapes than Wynn. Pete has 7 grapes. How many grapes does Wynn have? Show your thinking.

_____ grapes

On My Own

Name _____

What equation can show the problem? Use ? for the unknown. Then solve.

1. Lin has 3 more books than Malik. Malik has 6 books. How many books does Lin have?

   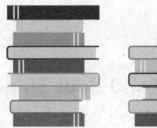
   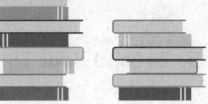

   Equation:

   _____ books

2. Kai has 2 fewer markers than Oscar. Kai has 9 markers. How many markers does Oscar have?

   Equation:

   _____ markers

3. Maddie's class has 2 fewer computers than Logan's class. Maddie's class has 7 computers. How many computers does Logan's class have?

   Equation:

   _____ computers

4. **Error Analysis** There are 3 fewer pennies than dimes. There are 8 pennies. How many dimes?

Alexis writes 8 − 3 = ? to show the number of dimes. Do you agree? Explain.

5. **Extend Your Thinking** How can you write a problem to match the picture? Make the greater number unknown.

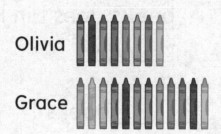

Olivia

Grace

Reflect

How do you solve a compare problem when the greater number is unknown?

Math is... Mindset
What did you do to be ready to learn today?

Showing Addition and Subtraction

Name _____

1. There are 8 red balloons and 3 blue balloons. How many more red balloons are there than blue balloons?

   Circle the equation that shows the problem.

   8 + 3 = _____   OR

   8 − 3 = _____

   Tell or show why.

---

2. Sam has 11 fish in his tank. Sue has 4 fewer fish in her tank. How many fish does Sue have?

   Circle the equation that shows the problem.

   11 + 4 = _____   OR

   11 − 4 = _____

   Tell or show why.

3. There are 7 rabbits in the field. If 5 more rabbits hop into the field, how many rabbits are now in the field?

Circle the equation that shows the problem.

7 + 5 = _____   OR

7 − 5 = _____

Tell or show why.

Reflect On Your Learning

Represent and Solve More Compare Problems

? Be Curious

Mia buys more bananas than Carter.

Mia buys some bananas.

How many bananas does Carter buy?

Math is... **Mindset**

What helps you understand your classmates' ideas?

Learn

Mia has 2 more bananas than Kenji.
Mia has 7 bananas.

How many bananas does Kenji have?

Cubes can show the problem.

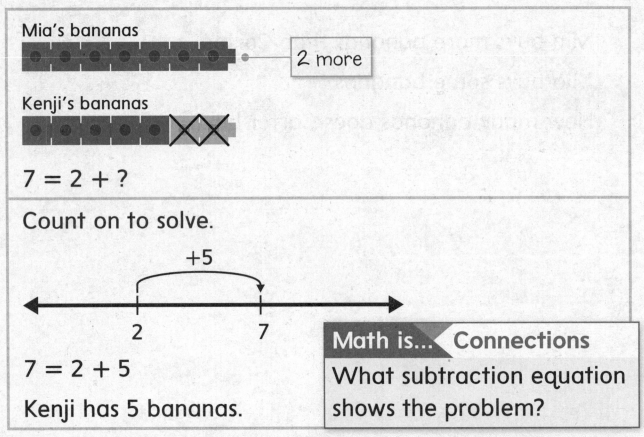

Mia's bananas

2 more

Kenji's bananas

$7 = 2 + ?$

Count on to solve.

+5

2   7

$7 = 2 + 5$

Kenji has 5 bananas.

Math is... Connections

What subtraction equation shows the problem?

You can add or subtract to find the lesser number.

💬 Work Together

Owen eats 4 more strawberries than Bella. Owen eats 12 strawberries. How many strawberries does Bella eat?

Show your thinking.

_____ strawberries

On My Own

Name _____

What equation can show the problem? Write an equation. Use ? for the unknown. Then solve.

1. Jack has 3 fewer pumpkins than Lily. Lily has 5 pumpkins. How many pumpkins does Jack have?
Equation:

_____ pumpkins

2. Gabby picks 3 fewer peaches than Mason. Mason picks 8 peaches. How many peaches does Gabby pick?
Equation:

_____ peaches

3. Brooklyn eats 2 more carrots than Aubrey. Brooklyn eats 13 carrots. How many does Aubrey eat?
Equation:

_____ carrots

4. **STEM Connection** Ruby gives the cats 6 fewer treats than the dogs. She gives the dogs 16 treats. How many treats does Ruby give the cats?

Draw to show your thinking.

_____ treats

5. **Extend Your Thinking** Write or draw problem with an unknown. Then solve.

🐾 **Reflect**

How can you add or subtract to find how many more or how many fewer?

Math is... **Mindset**

How have you shown that you understand your classmates' ideas?

Solve Compare Problems Using Addition and Subtraction

Be Curious

What could the question be?

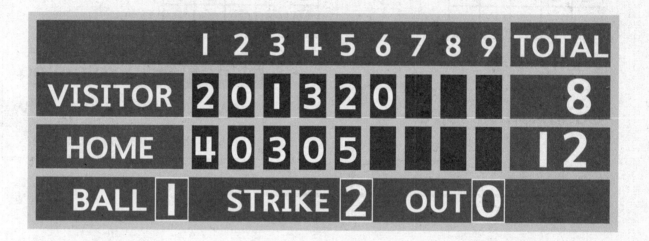

	1	2	3	4	5	6	7	8	9	TOTAL
VISITOR	2	0	1	3	2	0				8
HOME	4	0	3	0	5					12

BALL 1   STRIKE 2   OUT 0

Math is... Mindset

What about math makes you feel most confident?

Learn

There are 7 fewer basketballs than baseballs. There are 12 basketballs.

How many baseballs are there?

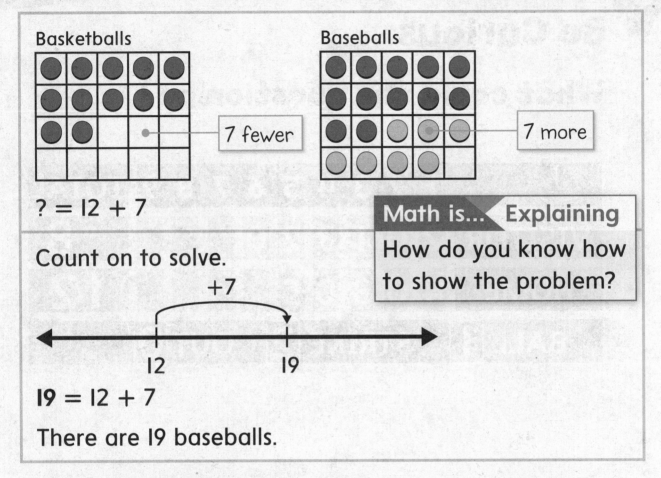

Basketballs

Baseballs

7 fewer

7 more

? = 12 + 7

Count on to solve.

+7

12   19

19 = 12 + 7

There are 19 baseballs.

Math is... Explaining

How do you know how to show the problem?

You can add or subtract to find how many more or how many fewer.

Work Together

Kylie stacks 13 blocks. Ella stacks 19 blocks. How many fewer blocks did Kylie stack than Ella?

Show your thinking.

_____ blocks

On My Own

Name _____

What equation can show the problem? Write an equation. Use ? for the unknown. Then solve.

1. Sam has 6 more puppets than Laura. Laura has 7 puppets. How many puppets does Sam have?

   Equation:

   _____ puppets

2. Jackson has 6 fewer berries than Tammy. Tammy has 16 berries. How many berries does Jackson have?

   Equation:

   _____ berries

3. Ava has 8 stuffed animals. Carston has 14 stuffed animals. How many more stuffed animals does Carston have than Ava?

   Equation:

   _____ stuffed animals

4. Error Analysis There are 5 forks and 7 spoons. How many fewer forks are there?

Sandra says there are 12 fewer forks. Do you agree with Sandra? Explain.

5. Extend Your Thinking Write or draw a problem to match the equation $? + 6 = 14$. Use the word *more*.

↩ Reflect

How do you know whether to add or subtract to solve a how many more or how many fewer problem?

Math is... Mindset

What helped you feel confident doing math?

Unit Review

Name _____

Vocabulary Review

Use the vocabulary to complete the sentence.

addend	compare
equation	unknown
problem	

1. You can _____ numbers to find the difference.

2. The _____ is a missing number in an equation.

3. Any number being added to another number is an _____.

4. A _____ is a question to answer.

5. A number sentence that includes the equal sign is an _____.

Review

6. Heather has 5 cubes. Amir has 14 cubes. How many fewer cubes does Heather have than Amir? Which equations match the word problem? **Choose all the correct answers.**

Heather ●●●●●

Amir ●●●●●●●●●●●●●●

A. $14 - 5 = 9$

B. $5 + 14 = 19$

C. $5 + 9 = 14$

7. Liz buys 5 more apples than oranges. Liz buys 7 oranges. How many apples does Liz buy?

A. 2 apples   B. 6 apples

C. 9 apples   D. 12 apples

8. Hector has 3 fewer nails than Gio. Gio has 11 nails. How many nails does Hector have?
Make an equation to show the problem.
Use ? for the unknown. Then solve.

Equation:

_____ nails

9. Luke finds 8 more shells than Flora. Luke finds 15 shells. How many shells does Flora find?

_____ shells

10. Jodi sees 16 leaves. Keith sees 12 leaves. How many fewer leaves does Keith see than Jodi?

_____ leaves

11. Mack plants 14 trees. Cindy plants 19 trees. How many more trees does Cindy plant than Mack? Draw to show your thinking.

_____ trees

12. Sam has 5 fewer rings than Alyssa. Sam has 8 rings. How many rings does Alyssa have? Make an equation to show the problem. Use ? for the unknown. Then solve.

Equation:

_____ rings

Performance Task

A museum has more pictures of airplanes than pictures of rockets. There are 17 pictures in all.

Part A: How might you show the number of airplane pictures and the number of rocket pictures?

Part B: Could there be 8 pictures of airplanes? Tell how you know.

Part C: 3 more rocket pictures get donated to the museum. There are still more airplane pictures than rocket pictures. How many airplane pictures could there be?

Reflect

How can you use addition and subtraction to compare?

Fluency Practice

Name _____

Fluency Strategy

You can use doubles to subtract.

$6 - 3 = ?$

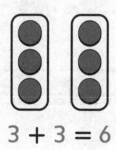

$3 + 3 = 6$

You know $3 + 3 = 6$.

So, $6 - 3 = 3$.

1. How can you use doubles to subtract $8 - 4$?

Fluency Flash

What is the difference? Write a double to help you subtract.

2. $10 - 5 =$ _____

3. $4 - 2 =$ _____

_____   _____

Fluency Check

What is the sum or difference? Write the number.

4. $4 + 4 =$ _____

5. $2 + 0 =$ _____

6. $5 + 5 =$ _____

7. $3 + 0 =$ _____

8. $7 - 0 =$ _____

9. $8 - 4 =$ _____

10. $2 - 1 =$ _____

11. $8 - 0 =$ _____

12. $12 - 6 =$ _____

13. $2 + 2 =$ _____

Fluency Talk

How does knowing a doubles fact help you subtract?

How can you show $6 + 6 = 12$? Explain your work.

Subtraction within 100

Focus Question

What strategies help me subtract 2-digit numbers?

Hi, I'm Hugo!

I want to be a meteorologist. Meteorologists study the weather. They use subtraction to tell changes in temperature.

Weather **TODAY**

85°F
Noon

STORM SAFETY

STEM video | GO ONLINE

141

Name _____

Put It All Together

Find the missing numbers.

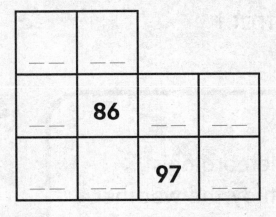

Copyright © McGraw-Hill Education

Be Curious

What do you see?

1	2	3	4	5	6	7	8	9	10
11	12	13	14	15	16	17	18	19	20
21	22	23	24	25	26	27	28	29	30
31	32	33	34	35	36	37	38	39	40
41	42	43	44	45	46	47	48	49	50
51	52	53	54	55	56	57	58	59	60
61	62	63	64	65	66	67	68	69	70
71	72	73	74	75	76	77	78	79	80
81	82	83	84	85	86	87	88	89	90
91	92	93	94	95	96	97	98	99	100

Math is... Mindset

What makes you feel excited in math?

Learn

The bookcase has 73 books.
The librarian takes 10 books.

How many books are there now?

You can use mental math to subtract 10 from a
2-digit number.

$73 - 10 = 63$

$56 - 10 = 46$

$82 - 10 = 72$

> **Math is...** Patterns
>
> What patterns do
> you see?

When you subtract 10, the tens digit goes down by 1
and the ones digit stays the same.

Work Together

What is $48 - 10$? Explain your thinking.

$48 - 10 = $ _____

On My Own

Name _____

Is the equation true? Circle Yes or No.

1. $28 - 10 = 18$

   Yes   No

2. $77 - 10 = 66$

   Yes   No

What is the difference?

3. $56 - 10 = $ _____

4. $79 - 10 = $ _____

5. $30 - 10 = $ _____

6. $64 - 10 = $ _____

7. $83 - 10 = $ _____

8. $95 - 10 = $ _____

9. Anya has 44 balloons. She gives her friend 10 balloons. How many balloons does Anya have?

_____ balloons

10. **Error Analysis** Toby reads 75 pages. Ethan reads 10 fewer pages than Toby. Ethan says he read 64 pages.
Do you agree with Ethan? Explain.

11. **Extend Your Thinking** How can you use mental math to subtract 45 − 10?

Reflect

How can you use mental math to subtract 10?

Math is... Mindset
What made you feel excited about math?

Represent Subtracting Tens

❓ Be Curious

Tell me everything you can.

Math is... **Mindset**

How do you help make everyone feel safe in class?

Learn

A store has 60 table tennis balls.
A shopper buys 20 table tennis balls.

How many table tennis balls are left?

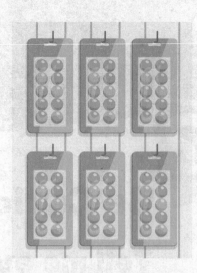

$60 - 20 = ?$

Use blocks to show 60.	Cross out 2 tens to subtract 20.
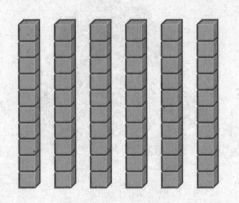	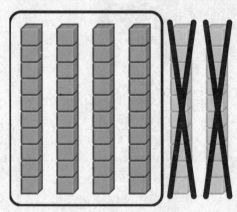
6 tens = 60	4 tens remain.
	$60 - 20 = \mathbf{40}$

You can show subtracting tens with blocks.

> **Math is...** Connections
>
> How can $6 - 2 = 4$ help you subtract $60 - 20$?

Work Together

What is the difference? Show your thinking.

$90 - 40 =$ _____

On My Own

Name _____

**What subtraction equation matches the tens blocks?
Write the equation.**

1.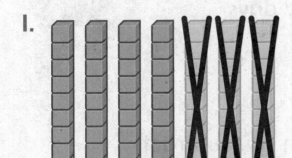

2.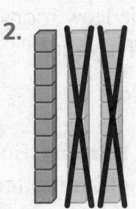

What is the difference? Show your thinking.

3. $50 - 40 =$ _____

4. $60 - 30 =$ _____

5. $90 - 70 =$ _____

6. $80 - 10 =$ _____

7. There are 70 muffins at the bakery. The baker sells 50 muffins. How many muffins are left?

_____ muffins

8. **STEM Connection** There were 80 rainy days last year. There were 30 fewer rainy days this year. How many rainy days were there this year?

_____ rainy days

9. **Extend Your Thinking** How can you subtract 40 – 10 without using tens blocks? Show or explain your thinking.

40 – 10 = _____

 **Reflect**

How can you show subtracting tens?

Math is... Mindset
How did you help everyone feel safe in class?

Subtract Tens

? Be Curious

What do you notice?
What do you wonder?

Math is... Mindset

How do you understand the thinking that is different from yours?

Learn

How much greater is
Dave's number than Amanda's?

Dave	Amanda
80	30

▶ **One Way** Use a number chart.

41	42	43	44	45	46	47	48	49	50
51	52	53	54	55	56	57	58	59	60
61	62	63	64	65	66	67	68	69	70
71	72	73	74	75	76	77	78	79	80

Subtract 30.

Start at 80.

$80 - 30 = 50$

▶ **Another Way** Use an open number line.

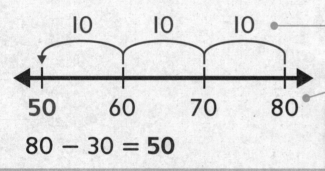

Subtract 30.

Start at 80.

$80 - 30 = 50$

Math is... Structure

Why can you count back
by tens instead of ones?

When you subtract tens from
2-digit numbers, only the digits in the tens place change.

💬 Work Together

What is the difference? Show your thinking.

$70 - 50 = $ _____

On My Own

MATH REPLAY | GO ONLINE

Name _____

What is the difference? Use the number chart.

1. $80 - 20 =$ _____

2. $60 - 40 =$ _____

3. $70 - 10 =$ _____

4. $70 - 70 =$ _____

1	2	3	4	5	6	7	8	9	10
11	12	13	14	15	16	17	18	19	20
21	22	23	24	25	26	27	28	29	30
31	32	33	34	35	36	37	38	39	40
41	42	43	44	45	46	47	48	49	50
51	52	53	54	55	56	57	58	59	60
61	62	63	64	65	66	67	68	69	70
71	72	73	74	75	76	77	78	79	80
81	82	83	84	85	86	87	88	89	90
91	92	93	94	95	96	97	98	99	100

What is the difference? Use the open number line.

5. $50 - 20 =$ _____

6. $90 - 40 =$ _____

7. T.J. has 40 baseball cards. He gives away 20 cards. How many cards does T.J. have left?

_____ baseball cards

8. **Error Analysis** Maya uses an open number line to show 80 − 50.

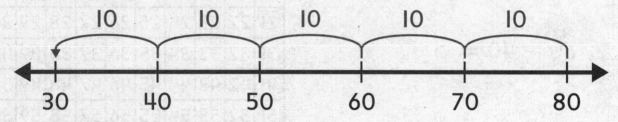

How do you respond to Maya?

9. **Extend Your Thinking** How can you use a number chart or number line to subtract 100 − 30?

⟳ **Reflect**

Which tool makes subtracting tens easier for you: a number chart or an open number line? Tell why.

Math is... **Mindset**
How did the thinking that was different from yours help you?

Be Curious

Which doesn't belong?

$$90 - 20 = ?$$

$$? = 20 + 70$$

$$70 + ? = 90$$

$$70 + 20 = ?$$

Math is... Mindset

What helps you stay focused on your work?

Learn

90 tennis balls are in a basket.
Briana uses 20 tennis balls to practice.

How many tennis balls are left?

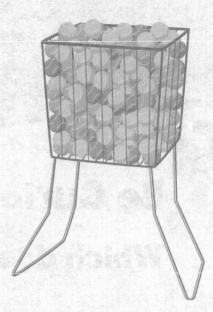

You can use addition to subtract tens.

$90 - 20 = ?$

Make an unknown addend equation. $20 + ? = 90$

Use a known fact. $2 + 7 = 9$

$20 + 70 = 90$, so $90 - 20 = 70$.

You can use addition and known facts to help you subtract tens.

Math is... Connections

How are the parts in the two equations related?

Work Together

How can you add to subtract $50 - 50$?
Show your thinking.

$50 - 50 = $ _____

On My Own

Name _____

What addition equation can you use to help you subtract? Write the equation.

1. $90 - 80 = ?$

2. $60 - 10 = ?$

How can you use addition to find the difference? Show your thinking.

3. $80 - 70 = $ _____

4. $50 - 10 = $ _____

5. $90 - 60 = $ _____

6. $30 - 30 = $ _____

7. Margo buys 70 stamps. She uses 20 stamps. How many stamps does Margo have left?

_____ stamps

8. **STEM Connection** Emily reads about planes. She reads 90 minutes on Monday. She reads 30 minutes on Thursday. How many more minutes does Emily read on Monday than Thursday?

_____ minutes

9. **Extend Your Thinking** Make a word problem about subtracting tens from tens. Make an addition equation to help you solve it.

 **Reflect**

How can you use addition equations to subtract?

Math is... **Mindset**
What helped you stay focused on your work today?

Showing Problems with Tens

Name _____

1. Roshni has 43 flowers. She gives 10 flowers to her friend.

How many flowers does Roshni have left?

Show the problem using the number chart.

1	2	3	4	5	6	7	8	9	10
11	12	13	14	15	16	17	18	19	20
21	22	23	24	25	26	27	28	29	30
31	32	33	34	35	36	37	38	39	40
41	42	43	44	45	46	47	48	49	50
51	52	53	54	55	56	57	58	59	60
61	62	63	64	65	66	67	68	69	70
71	72	73	74	75	76	77	78	79	80
81	82	83	84	85	86	87	88	89	90
91	92	93	94	95	96	97	98	99	100

Circle the equation that shows the problem.

$43 - 10 =$ _____   OR $43 + 10 =$ _____

Tell or show why you chose that equation.

2. Lori has 60 stickers. She has 30 fewer stickers than Cory. How many stickers does Cory have?

Show the problem using tens blocks.

Circle the equation that shows the problem.

60 − 30 = _____   OR   60 + 30 = _____

Tell or show why you chose that equation.

Reflect On Your Learning

Be Curious

What could the question be?

Math is... Mindset

What do you do to be an active listener?

Learn

Leah has 80 wax sticks.
Brooks has 60 wax sticks.

How many more wax sticks does Leah have than Brooks?

$\underline{8}$ tens $- \underline{6}$ tens $= \underline{2}$ tens

$80 - 60 = 20$

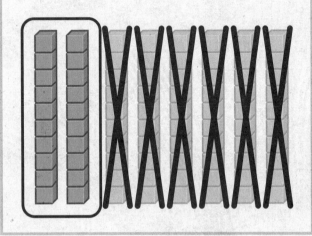

$60 + ? = 80$

$6 + 2 = 8$

$60 + 20 = 80,$

so $80 - 60 = 20$.

Math is... Explaining

How did you use place value to subtract $80 - 60$?

You can use different strategies to subtract tens.

Work Together

How can you subtract $70 - 40$? Explain your thinking.

On My Own

Name _____

Is the equation true? Circle Yes or No. Explain how you know.

1. 10 − 10 = 0

   Yes   No

2. 90 − 50 = 30

   Yes   No

What is the difference? Explain or show your thinking.

3. 67 − 10 = _____

4. 90 − 20 = _____

5. 70 − 60 = _____

6. A parking lot has 80 cars. Then 80 cars drive away. How many cars are still in the parking lot?

_____ cars

7. **Error Analysis** Yolanda draws to subtract 40 − 20. She says the difference is 30. Do you agree? Explain.

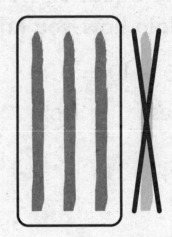

8. **Extend Your Thinking** How can you subtract 100 − 80?

Reflect

Which strategy do you find works best for you when subtracting tens from tens? Why?

Math is... Mindset

What helped you be an active listener today?

Unit Review

Vocabulary Review

Use the vocabulary to complete the sentence.

difference	ones
digit	place value
equation	tens

1. A symbol used to write a number is a

   _____.

2. Every digit in a number has _____.

3. In the subtraction equation $18 - 4 = 14$, 14 is

   the _____.

4. The number 85 has 8 _____.

5. A number sentence that has an equal sign is an

   _____.

6. The number 64 has 4 _____.

Review

7. A store has 80 watches. It sells 10 watches.
   How many watches does the store have now?

   A. 70 watches   B. 79 watches

   C. 81 watches   D. 90 watches

8. Use the tens blocks to subtract.

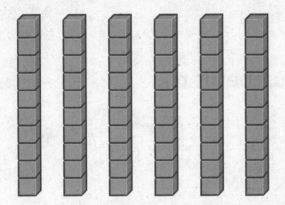

   60 students are on the playground.
   20 students leave. How many
   students stay on the playground?

   A. 30 students   B. 40 students

   C. 50 students   D. 60 students

9. How can you use addition to find the difference?
   Write the addition equation.

   $70 - 30 = $ _____

10. There are 50 markers in a classroom. 30 markers are black and the rest are blue. How many markers are blue? Use the number line to subtract.

_____ blue markers

11. Roderick bakes 70 muffins. He sells 30 muffins. How many muffins does Roderick have left? Use the number chart.

_____ muffins

1	2	3	4	5	6	7	8	9	10
11	12	13	14	15	16	17	18	19	20
21	22	23	24	25	26	27	28	29	30
31	32	33	34	35	36	37	38	39	40
41	42	43	44	45	46	47	48	49	50
51	52	53	54	55	56	57	58	59	60
61	62	63	64	65	66	67	68	69	70
71	72	73	74	75	76	77	78	79	80
81	82	83	84	85	86	87	88	89	90
91	92	93	94	95	96	97	98	99	100

12. How can you subtract $90 - 30$? Show or tell how you subtracted.

$90 - 30 =$ _____

Performance Task

Hugo has been recording the weather. He records 20 more cloudy days than rainy days. He records 10 more sunny days than cloudy days.

Part A: Did Hugo record more sunny days than rainy days? Tell how you know.

Part B: If Hugo recorded 50 sunny days, what can you say about the number of cloudy and rainy days Hugo recorded?

 **Reflect**

How did you use different ways to subtract tens?

Fluency Practice

Name _____

Fluency Strategy

You can use doubles to help you add.

$3 + 4 = ?$

What double can I use to add $3 + 4$?

$3 + 3 = 6$

Then add 1 more.

So, $3 + 4 = 7$

1. How can you use doubles to add $4 + 6$?

Fluency Flash

What is the sum? Use doubles to help you add.

2. $4 + 5 =$ _____

3. $1 + 3 =$ _____

Fluency Check

What is the sum or difference? Write the number.

4. $2 + 4 =$ _____

5. $2 + 3 =$ _____

6. $3 + 3 =$ _____

7. $1 + 2 =$ _____

8. $10 - 5 =$ _____

9. $2 + 2 =$ _____

10. $8 - 4 =$ _____

11. $2 - 1 =$ _____

12. $1 + 1 =$ _____

13. $4 + 4 =$ _____

Fluency Talk

How can you use doubles to add $3 + 5$?
Explain your work.

How can you show $6 - 3 = 3$? Explain your work.

Measurement and Data

Focus Question

How can I use tools to measure and interpret data?

Hi, I'm Deven.

I want to be a sound engineer. What's your favorite instrument? That's the question I asked my friends. I wonder how I can show their answers.

STEM video | GO ONLINE

Name _____

How Long Can You Build It?

Look at the trains. What do you notice?

Compare and Order Lengths

Be Curious

Which doesn't belong?

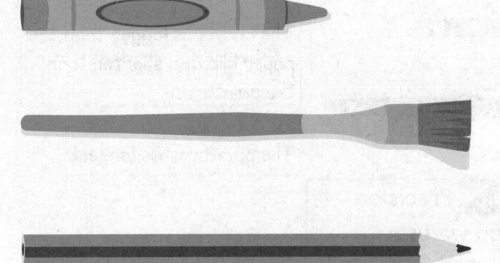

Math is... Mindset

What helps you stay focused on your work?

Learn

Declan wants to put these objects in order from shortest to longest.

How can you order these objects?

You line up the ends to compare **lengths** of objects.

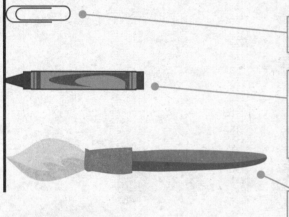

The paper clip is **shortest**.

The crayon is **longer** than the paper clip and **shorter** than the paintbrush.

The paintbrush is **longest**.

Math is... Precision

When do we use "longer" and "longest"?

Work Together

Circle the longest object. Put an X on the shortest object.

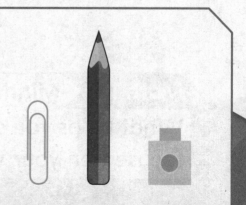

On My Own

Name _____

1. Which instrument is shorter? Choose the correct answer.

A.

B.

Compare and order the objects.
Write A, B, or C.

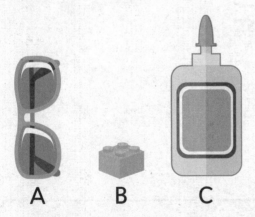

A   B   C

2. Object _____ is shortest.

3. Object _____ is longest.

4. Object A is shorter than object _____.

5. Object A is longer than object _____.

Circle the longest object. Put an X on the shortest object.

6.

7.

8. **Extend Your Thinking** Find 4 objects. Order the objects from longest to shortest.

 **Reflect**

How can you compare and order lengths?

Math is... Mindset

What helped you stay focused on your work today?

More Ways to Compare Lengths

Be Curious

What question could you ask?

Math is... Mindset

How can your strengths help you learn today?

Learn

How can you use the spatula to compare the spoon and the rolling pin?

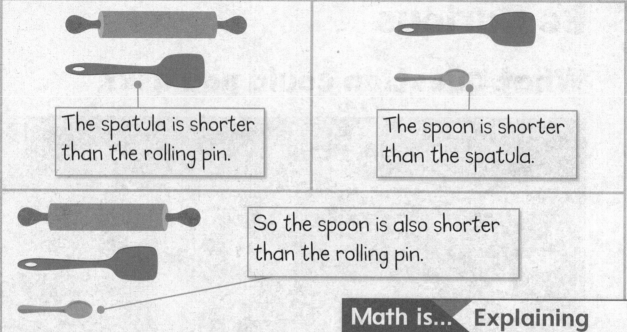

The spatula is shorter than the rolling pin.

The spoon is shorter than the spatula.

So the spoon is also shorter than the rolling pin.

You can compare the lengths of two objects using a third object.

Math is... Explaining

Could you use the rolling pin to compare the spatula and the spoon?

💬 Work Together

Is the pencil shorter or longer than the marker? Explain your thinking.

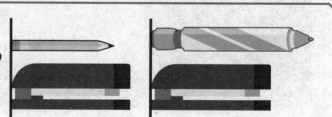

The pencil is _____ than the marker.

On My Own

Name _____

Compare the lengths of two objects using a third object. Write *longer* or *shorter*.

1. The phone is _____ than the flashlight.

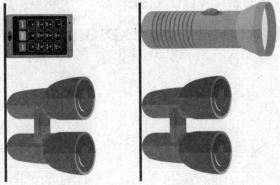

2. The shoe is _____ than the bell.

3. The dog treat is _____ than the toothbrush.

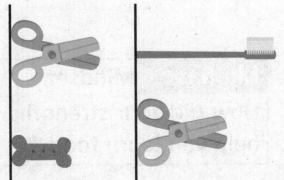

4. **STEM Connection** How can Jin use bone B to compare the lengths of bone A and bone C? Explain.

A

B

C

5. **Extend Your Thinking** Draw 3 lines that are different lengths. Which line can you use to compare the lengths of the other two? Explain.

⊘ Reflect

How can you use an object to compare the lengths of two other objects?

Math is... Mindset

How did your strengths help you learn today?

Strategies to Measure Lengths

Be Curious

What do you notice?
What do you wonder?

Math is... Mindset
What are your strengths in math?

Learn

How can you tell the length of the book?

You use **units** that are the same size to **measure** length.

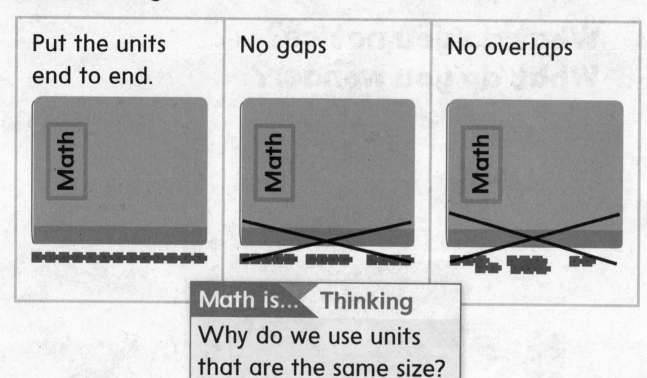

| Put the units end to end. | No gaps | No overlaps |

Math is... Thinking

Why do we use units that are the same size?

Work Together

How many paper clips long is the hot dog?

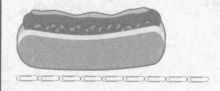

_____ paper clips long

On My Own

Name _____

1. How many units long is the crayon?

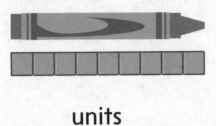

_____ units

2. How many nickels long is the pencil?

_____ nickels

3. How many blocks long is the comb?

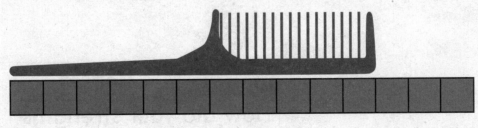

_____ blocks

4. Error Analysis Lydia says the hammer is 7 connecting cubes long.

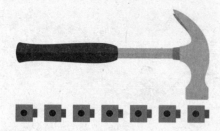

Do you agree with Lydia? Explain.

5. Extend Your Thinking How long is your desk? Choose a unit and measure the length of your desk. Explain how you measured.

My desk is _____ _____ long.

 **Reflect**

How can you measure length using units?

Math is... **Mindset**
How did your strengths in math help you today?

How Long Is the Rope?

CHERYL TOBEY
MATH
PROBES

Name _____

Circle Yes or No to answer the question.

1. Is the rope 3 crayons long?

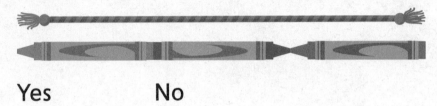

Yes   No

2. Is the rope 4 paper clips long?

Yes   No

3. Is the rope 4 cubes long?

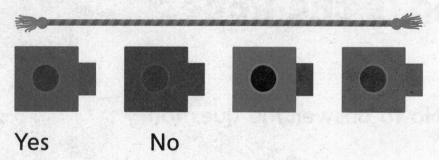

Yes   No

4. Is the rope 3 paper clips long?

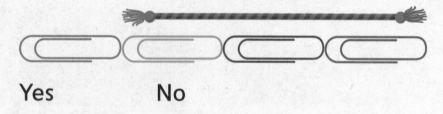

Yes   No

Reflect On Your Learning

More Strategies to Measure Lengths

Be Curious

Tell me everything you can.

Math is... Mindset
What helps you be ready to learn?

Learn

Yao and Maya measure the shovel.
They use different units.

How will their answers compare?

Yao uses cubes.

The shovel is 8 cubes long.

Maya uses paper clips.

The shovel is 5 paper clips long.

Their answers are different.
They use different units.

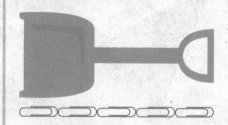

Units of different sizes give
different measurements.

Math is... Connections

Which unit is larger?

 Work Together

Will you use more connecting cubes
or paper clips to measure the length
of the shoe?

On My Own

Name _____

1. How long is the toy truck?

   _____ golf balls

   _____ staplers

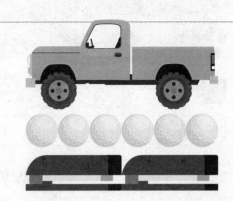

2. How long is the toy boat?

   _____ tape holders

   _____ erasers

How can you use two different units to measure the toothpaste tube?

3. Which unit is smaller?

   A. dimes   **B.** squares

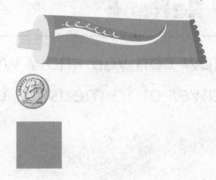

4. Will you use *more* or *fewer*
   dimes than squares to measure
   the toothpaste?

   _____ dimes than squares

5. Will you use *more* or *fewer* keys than connecting cubes to measure the sandal? Explain.

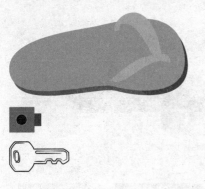

_____ keys than connecting cubes

6. **Extend Your Thinking** Choose two units to use to measure the length of your book. Which unit will you use more of to measure your book? Explain.

Reflect

How can you know which unit you will use *more* or *fewer* of to measure length?

Math is... **Mindset**
What helped you be ready to learn?

Tell Time to the Hour

Be Curious

How are they the same?
How are they different?

Math is... Mindset

How do you show that you respect your classmates?

Learn

How can you tell what time it is?

Math is... Explaining

Why is it important to know how to tell time?

An **analog clock** shows the **hour** and **minutes** with hands.

minute hand

hour hand

This clock shows 2 o'clock.

A **digital clock** shows the hour and minutes with numbers.

minutes

hour

This clock also shows 2 o'clock.

You can use clocks to tell and write time to the hour.

Work Together

What time is it? Explain how you know.

_____ : _____

On My Own

Name _____

What time is it? Write the time.

1.

_____ : _____

2.

11:00

_____ : _____

3.

_____ : _____

4.

_____ : _____

5.

5:00

_____ : _____

6.

_____ : _____

7. **STEM Connection** Deven listens to music at the time on the clock. What time is it?

_____ : _____

8. Football practice starts at 3 o'clock. What time does it start?

_____ : _____

9. **Extend Your Thinking** How can you show 4:00 on the clock? Draw the hour hand and minute hand.

Reflect

How can you tell time with a clock that has hands?

Math is... Mindset
How did you show respect to your classmates?

Tell Time to the Half Hour

? Be Curious

What do you notice?
What do you wonder?

Math is... Mindset
What helps you understand how others are feeling?

Learn

What time is it?

Why do you think the hour hand is between 1 and 2?

The hour hand is between 1 and 2, so the hour is 1.

The minute hand points to 6, so the minutes are 30.

The clock shows 1:30 or half past 1.

1 hour is 60 minutes

This clock also shows 1:30 or half past 1.

You can use clocks to tell and write time to the **half hour**.

Work Together

What time is it? Draw the hour and minute hands to show the same time.

On My Own

Name _____

What time is it? Write the time.

1.

_____ : _____

2.

_____ : _____

3.

_____ : _____

4.

_____ : _____

5.

_____ : _____

6.

_____ : _____

7. Error Analysis Sally says the time is 3:30. Steve says the time is half past 3:00. How can they both be correct?

8. Extend Your Thinking How can you show half past 9:00? Draw the hour and minute hands.

 **Reflect**

How can you tell time to the half hour?

Math is... Mindset

What helped you understand how others were feeling?

Organize Data

? Be Curious

What do you see?

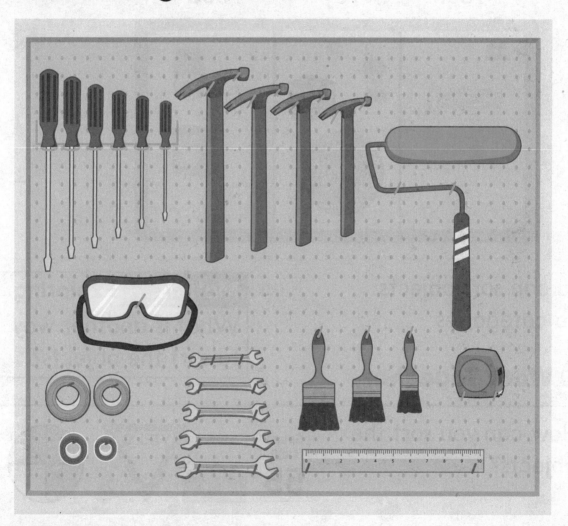

Math is... Mindset

How do you work well with a partner to complete a task together?

Learn

How can you sort the objects?

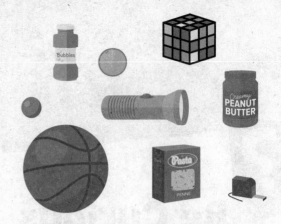

You can sort the objects by use.

Tool	Toy	Food

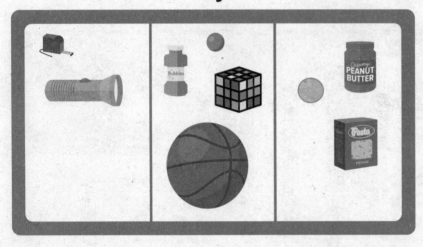

You can sort objects into categories.

Math is... Exploring

What is another way to sort the objects?

 Work Together

How can you sort the objects?

 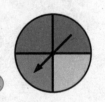

On My Own

Name _____

How can you sort the objects? Draw the objects or write their names to complete the chart.

chalk   apple   trash can   pea   box   pencil

1.

Small	Medium	Large

2.

Container	Writing Tool	Food

3. How can you sort these objects? Complete the chart. Write the category names above the chart.

lime   tennis   trumpet   helmet   cherry   drum
        ball

4. Extend Your Thinking List 5 objects in your classroom. How can you sort them? Explain.

 **Reflect**

How can you sort objects into categories?

Math is... Mindset

What helped you work well with a partner to complete a task together?

Represent Data

Be Curious

What do you notice?
What do you wonder?

Math is... Mindset

What are some ways you can contribute to your group today?

Learn

Jaime wants to know how many friends choose the butterfly as their favorite bug. **How can he collect his friends' choices?**

You can make a table. The table has one row for each choice.

Bug	Tally
🦋	
🐝	
🦗	

Math is... Explaining

Why should you make a tally mark for each bug?

Use I tally mark to show each choice.

Bug	Tally
🦋	ⅢⅢ I
🐝	III
🦗	ⅢⅢ

5 tally marks

Jaime can use tally marks to collect his friend's choices.

A tally chart shows how many of each kind of object.

Work Together

How can you show the data in the tally chart? Make tally marks and write numbers to complete the chart.

Sport	Tally	Total
🏈		
🏀		
⚽		

On My Own

Name _____

Use the tally chart to answer the questions.

1. How many of each kind of pet do students have? Write the totals in the chart.

2. How many pets are there in all?

   _____ pets

Pet	Tally	Total
🐟	IIII	
🐕	HHH III	
🐈	HHH I	

Use the data to complete the chart.

3. Students saw some bugs at the park. Make tally marks to complete the chart.

Bug	Tally
🐞	
🐜	
🐝	

Toy	Tally
	卌 I
	卌 IIII
	IIII

4. **Error Analysis** Jade says 15 students voted for their favorite outdoor toy. Do you agree with Jade? Explain.

5. **Extend Your Thinking** When you make a tally chart, how do you know if you recorded all of the data?

🔄 **Reflect**

How can you sort data into a tally chart?

Math is... Mindset

In what ways have you contributed to your group today?

Interpret Data

? Be Curious

How are they the same?
How are they different?

Bug	Tally
🦋	IIII I
🐝	III
🦗	IIII

Favorite Bug

Bug	Tally	Total
🦋	IIII I	6
🐝	III	3
🦗	IIII III	8

> **Math is...** Mindset
>
> What helps you make sense of a situation?

Learn

Michael makes a chart to show his friends' favorite colors.

Color	Tally
blue ▬	⁙⁙
green ▬	⁙⁙ l
orange ▬	lll

How many friends does Michael ask?

You can use a **tally chart** to organize the **data**.

You can write the totals for each color.

Color	Tally	Total
▬	⁙⁙	5
▬	⁙⁙ l	6
▬	lll	3

You can add the totals to know how many friends Michael asks.

$5 + 6 + 3 = ?$

$11 + 3 = 14$

Michael asks 14 friends.

Math is... Connections

What other questions can you ask about the data?

🗨 Work Together

How many of each toy does Sarah have?

Sarah's Toy	Tally	Total
🚂	ll	
✈	llll	
🚗	⁙⁙	

On My Own

Name _____

Students pick what they want to see at the museum. Use the tally chart to answer the questions.

Attraction	Tally
	IIII
	IIII III
	IIII

1. How many students want to see the eagle?

   _____ students

2. How many students want to see the insect?

   _____ students

3. How many students want to see the dinosaur?

   _____ students

4. How many students chose to see an attraction?

   _____ students

Use the tally chart to answer the questions.

Weather	Tally
☀️	卌 I
❄️	III
⛈️	II

5. **STEM Connection** Hugo asks some friends to pick their favorite weather. How many friends pick sunny weather?

_____ friends

6. **Extend Your Thinking** How many friends in all pick their favorite weather? Explain how you know.

_____ friends

🎨 **Reflect**

How can you use tally charts to answer questions?

Math is... Mindset

What helped you make sense of a situation?

Solve Problems Involving Data

? Be Curious

Students pick their favorite instrument.

Some pick drums. Some pick piano.
Some pick trumpet.

Math is... Mindset

How do deep breaths
help you work better?

Learn

Annisa asks students to choose their favorite instrument from drums, trumpet, and piano.

How many more students choose drums than trumpet?

Instrument	Tally
🥁 |卌 |||
🎹	卌
🎺	卌	

8 students choose drums.

8 tally marks

 卌 |||

6 students choose trumpet.

6 tally marks

🎺 卌 |

$8 - 6 = 2$

2 more students choose drums than trumpet.

Math is... Exploring

What other questions can you ask about the data?

You can use a tally chart to answer questions about *how many more* and *how many less*.

💬 Work Together

How many zebras and lions are at the zoo?

_____ zebras and lions

Zoo Animals		Tally
Zebra	🦓	|||
Elephant	🐘	卌
Lion	🦁	||

On My Own

Name _____

Use the tally chart to answer the questions.

Favorite Toy	Tally
Soccer ball 🎱	卌 I
Bike 🚲	卌 IIII
Doll 👧	卌 II

1. Which toy did the most students choose?

   _____

2. Which toy did the fewest students choose?

   _____

3. How many more students chose bike than doll?

   _____ students

4. How many fewer students chose soccer ball than bike?

   _____ students

5. How many students chose a favorite toy?

   _____ students

Use the tally chart to answer the questions.

Type of Mug	Tally
Stripes 	ⵗⵗ II
Dots	IIII
Plain	ⵗⵗ I

6. How many more mugs have stripes than dots?

_____ mugs

7. How many mugs are there in all?

_____ mugs

8. **Extend Your Thinking** To have more mugs with dots than stripes, how many more dot mugs should there be? Explain your thinking.

_____ mugs

Reflect

How can you use a tally chart to answer questions?

Math is... Mindset

How has deep breathing helped you work better?

Unit Review

Name _____

Vocabulary Review

Match the name with the picture. Write the letter on the line.

1. analog clock _____

A.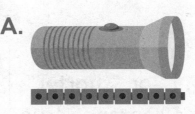

2. digital clock _____

B.

3. measure _____

C.

4. tally chart _____

D.
Sports We Play	Tally
⚾	IIII
⚽	IIII I
🎾	II

5. tally marks _____

E.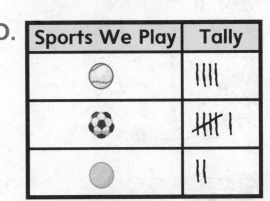

Review

6. Which animal sits on the shortest shelf?

   A. bear

   B. owl

   C. penguin

7. Complete the sentence with "longer" or "shorter" to make the sentence true.

   The ribbon with dots is _____ than the plain ribbon.

8. Which clock shows half past 4:00? Choose all the correct answers.

   A.

   B.

   C.

   D.

9. Look at the tally chart. How many more ants did they see than bees?

_____ more ants than bees

Insects We Saw	Tally
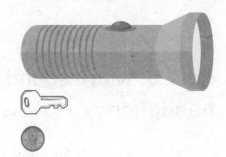	|||| ||||

10. Jay measures the length of a flashlight using keys and pennies. Will he use more keys or fewer keys than pennies to measure the flashlight? Write *more, fewer, larger,* or *smaller* to make the sentence true.

He will use _____ keys to measure the length of the flashlight because keys are a _____ unit.

11. How can you sort these objects? Write two or more categories.

Performance Task

Some sound engineers pick their favorite headphones.

Part A: How many pick each type of headphones? Make tally marks. Write the totals in the tally chart.

Favorite Headphones		
Type	**Tally**	**Total**

Part B: More sound engineers picked their favorite headphones. How would the data change? Explain.

Reflect

How can you use tools to measure and to answer questions about data?

Fluency Practice

Name _____

Fluency Strategy

You can use doubles to help you subtract.

$9 - 4 = ?$

What double can I use to subtract $9 - 4$?

I know $4 + 4 = 8$, so I know $8 - 4 = 4$.

9 is 1 more than 8.

Add 1 to the difference in $8 - 4 = 4$.

So, $9 - 4 = 5$

1. How can you use doubles to subtract $8 - 3$?

Fluency Flash

What is the difference? Use doubles to help you subtract.

2. $7 - 4$

I know _____ + _____ = _____.

So, I know _____ − _____ = _____.

7 is 1 less than _____.

Subtract _____ from the difference.

So, $7 - 4 =$ _____.

Fluency Check

What is the sum or difference? Write the number.

3. $9 - 5 =$ _____

4. $3 + 5 =$ _____

5. $4 + 6 =$ _____

6. $7 - 3 =$ _____

7. $6 - 2 =$ _____

8. $4 - 2 =$ _____

9. $8 - 4 =$ _____

10. $6 - 4 =$ _____

11. $6 - 3 =$ _____

12. $4 + 5 =$ _____

13. $2 + 3 =$ _____

14. $8 - 5 =$ _____

Fluency Talk

How can you use doubles to subtract $5 - 3$?

How can you show $4 + 5 = 9$?

Equal Shares

Focus Question

What are equal shares?

Hi, I'm Kayla.

Someday I want to be a landscape architect. Today I am planting my vegetable garden. I make my garden into 2 equal parts. I plant tomatoes in one part and cucumbers in the other.

Name _____

Cutting Squares in Half

Look at each picture. Choose one that is different from the others.

Picture A

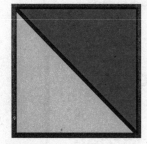

Picture B

Picture C

Picture D

Shade I half of each square in one color. Shade I half in another color. Shade the squares different ways.

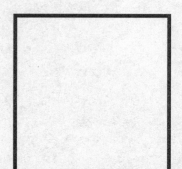

Understand Equal Shares

? Be Curious

Tell me everything you can.

Math is... Mindset
What do you do well
in math? In reading?

Learn

Kevin says the window in his room has 2 equal parts.

How can you tell if the 2 parts are equal?

These shapes have **equal shares**.

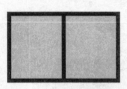

Math is... Modeling

What is another way to show a rectangle with equal squares?

These shapes do not have equal shares.

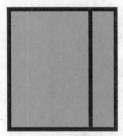

A **whole** has equal shares when the parts are the same size.

 Work Together

Which square has equal shares? Circle the square. Explain your thinking.

On My Own

Name _____

Does the shape have equal shares? Circle Yes or No.

1.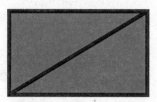

Yes   No

2.

Yes   No

3.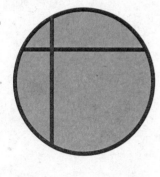

Yes   No

4.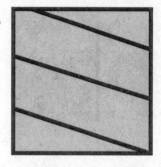

Yes   No

5.

Yes   No

6.

Yes   No

7. **STEM Connection** Does Kayla's garden have equal shares? Explain how you know.

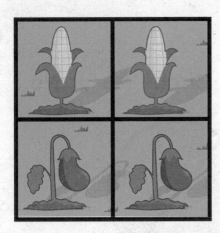

8. **Extend Your Thinking** Andrew and Jane share a snack bar. Do Andrew and Jane get equal shares of the snack bar? Explain how you know.

Reflect

How can you describe equal shares?

Math is... Mindset
What did you do well in math? In reading?

Partition Shapes into Halves

? Be Curious

What do you see?

Math is... Mindset

What helps you feel relaxed when you are frustrated?

Learn

Hieu folds his paper to make
2 equal shares.

How can you describe the parts?

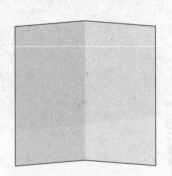

A whole with 2 equal shares shows **halves**.

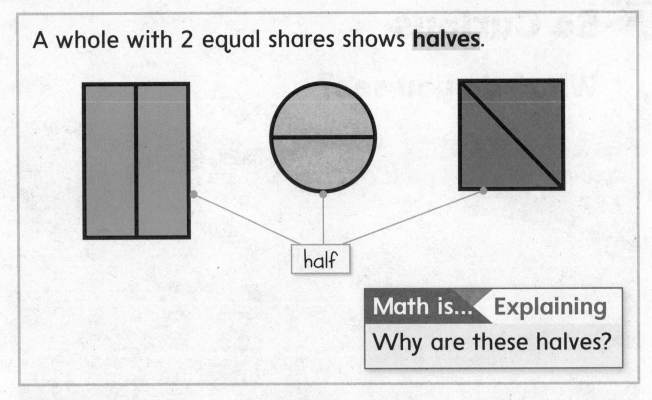

half

Math is... Explaining

Why are these halves?

Each equal share is **half of** the whole.

💬 Work Together

How can you make halves? Draw to show halves.
Explain your thinking.

On My Own

Name _____

Does the shape show halves? Circle Yes or No.

1.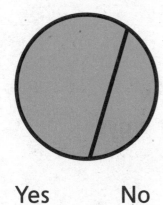

Yes   No

2.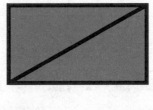

Yes   No

3. Which shapes show halves? Circle all the shapes.

How can you make halves? Draw to show halves.

4.

5.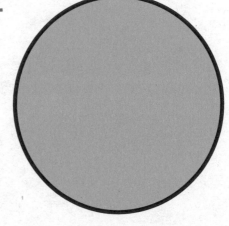

6. Error Analysis Jada says this circle shows halves. Do you agree with Jada? Explain your thinking.

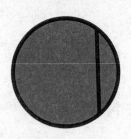

7. Extend Your Thinking How can you show halves of a square? Show two different ways. Label each half.

 **Reflect**

How can you know if a shape shows halves?

Math is... **Mindset**
What did you do to help you feel relaxed when you were frustrated?

Partition Shapes into Fourths

Be Curious

How are they the same?
How are they different?

Math is... Mindset

How do you show that you value your partner's ideas?

Learn

4 friends share a sandwich.

How can the friends make equal shares?

A whole with 4 equal shares shows **fourths**, or **quarters**.

fourth, or quarter

Each equal share is a **fourth of**, or a **quarter of**, the whole.

Math is... Connections

What is another meaning for *quarter*?

Work Together

How can you make fourths? Draw to show fourths. Explain your thinking.

On My Own

Name _____

Does the shape show fourths? Circle Yes or No.

1.

Yes   No

2.

Yes   No

3. **Which shapes show quarters? Circle all the shapes.**

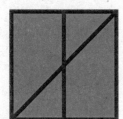   

How can you make fourths? Draw to show fourths.

4.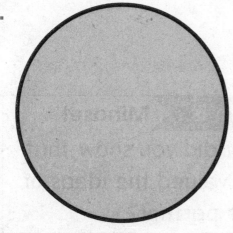

5.

6. **STEM Connection** Chloe puts some boards together. Do the boards show fourths? Explain your thinking.

7. **Extend Your Thinking** How can you show quarters of a rectangle? Show two different ways.

 **Reflect**

How can you know if a shape shows fourths?

Math is... Mindset

How did you show that you valued the ideas of your partner?

Partitioning into Fourths

CHERYL TOBEY
MATH
PROBES

Name

Decide if each notebook has been divided into fourths. Circle Yes or No.

1.

Yes   No

Tell or show how you know.

2.

Yes   No

Tell or show how you know.

3.

Yes   No

Tell or show how you know.

Decide if each notebook has been divided into fourths. Circle Yes or No.

4.

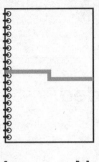

Yes　　　No

Tell or show how
you know.

5.

Yes　　No

Tell or show how
you know.

6.

Yes　　　No

Tell or show how
you know.

Reflect On Your Learning

Describe the Whole

Be Curious

What do you see?

Math is... Mindset

Why is it important to speak clearly and concisely?

Learn

Some friends share a pizza. The pizza is cut into 4 slices. The slices are the same size. Each student has 1 slice.

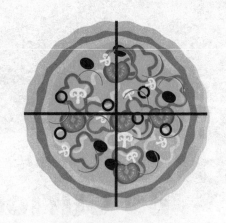

How many friends share the pizza?

A whole with 4 equal shares has 4 fourths.

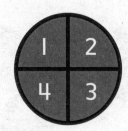

The pizza has 4 slices.

4 friends share the pizza equally.

Math is... Exploring

How is a whole with 4 equal shares different from a whole with 2 equal shares?

You can count the equal shares in a whole.

Work Together

How many quarters? Explain how you know.

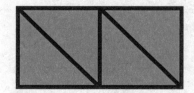

_____ quarters

On My Own

Name _____

1. Circle the shape that has 2 halves.

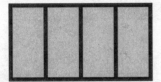 

2. Circle the shape that has 4 fourths.

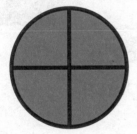 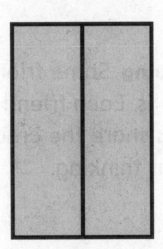

How many equal shares are there? Write the number.

3.

_____ quarters

4.

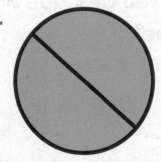

_____ halves

How many equal shares make up the whole?
Write *2 halves* or *4 fourths*.

5.

_____

6.

_____

7. **Error Analysis** Nia says there are 4 equal shares. Do you agree? Explain your thinking.

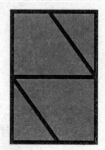

8. **Extend Your Thinking** Some friends break a cracker into fourths. Each friend gets 1 piece. How many friends share the cracker? Draw to show your thinking.

_____ friends

 Reflect

How can you describe how many equal shares make up a shape?

Math is... Mindset
How did speaking clearly and concisely help you today?

Describe Halves and Fourths of Shapes

Be Curious

How are they the same?
How are they different?

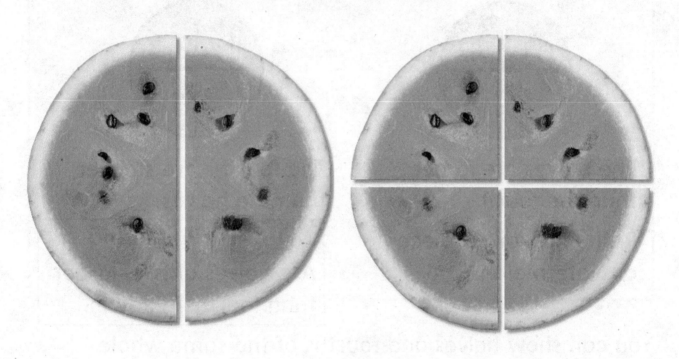

Math is... Mindset

What helps you make good decisions?

Learn

Reshu and Will cut their watermelon different ways.

Who has larger pieces of watermelon?

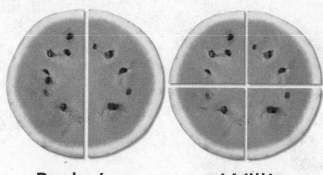

Reshu's
watermelon

Will's
watermelon

Reshu's watermelon is cut into 2 halves.

The halves are larger than the fourths.

Reshu has larger pieces of watermelon.

Will's watermelon is cut into 4 fourths.

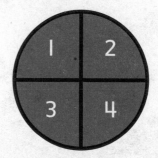

The fourths are smaller than the halves.

Math is... Exploring

Are halves always larger than fourths? Explain.

You can show halves and fourths of the same whole.

Work Together

Which shape has larger equal shares? Circle the shape. Explain your thinking.

On My Own

MATH REPLAY · GO ONLINE

Name _____

Circle the correct answer.

1. Which shows more equal shares?

   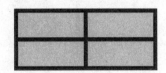

2. Which shows fewer equal shares?

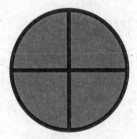   

3. Which shows larger equal shares?

4. Which shows smaller equal shares?

   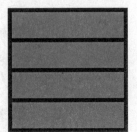

5. Which shows smaller equal shares? Circle the shape. Complete the sentence with *fewer* or *more* to make the sentence true.

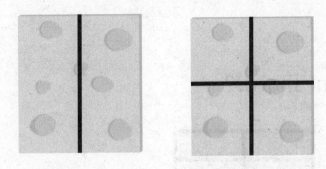

A whole with _____ equal shares has smaller shares.

6. Extend Your Thinking A circle shows 2 equal shares. Another circle that is the same size shows 4 equal shares. Which shares are larger? Explain.

Reflect

How are halves and fourths similar? How are they different?

Math is... Mindset
What helped you make good decisions today?

Unit Review

Name _____

Vocabulary Review

Use the vocabulary to complete each sentence.

equal shares	fourths
fourth of	half of
halves	whole

1. A whole with 2 equal shares shows _____.

2. Halves and fourths are different kinds

    of _____.

3. A whole with 4 equal shares shows _____.

4. When a shape shows fourths, each part is

    a _____ the whole.

5. When a shape shows halves, each part is

    a _____ the whole.

6. An entire shape is the _____.

Review

7. Which shapes show equal shares?
Choose all the correct answers.

A.

B.

C.

D.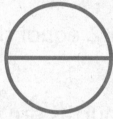

8. How many quarters does this shape show?

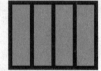 _____ quarters

9. Which shows smaller equal shares?
Circle the shape. Explain your thinking.

10. Which shapes show halves? Choose all the correct answers.

A.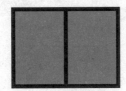

B.

C.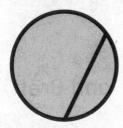

D.

11. Which shapes show fourths? Choose all the correct answers.

A.

B.

C.

D.

12. How many halves does this shape have?

_____ halves

Performance Task

Jessie and Brett each plant a garden. The gardens are the same size. Jessie marks her garden into fourths. Brett marks his garden into halves.

Part A: Draw to show Jessie's garden and Brett's garden.

Part B: How many equal shares do Jessie and Brett each make?

Part C: Who makes larger equal shares? Explain how you know.

 **Reflect**

What are equal shares?

Fluency Practice

Name _____

Fluency Strategy

You can use place value to help you add or subtract 10.

$4 + 10 = ?$

1	2	3	4	5	6	7	8	9	10
11	12	13	14	15	16	17	18	19	20
21	22	23	24	25	26	27	28	29	30

When you add 10, the tens digit goes up 1.

The ones digit is the same.

So, $4 + 10 = 14$.

$16 - 10 = ?$

1	2	3	4	5	6	7	8	9	10
11	12	13	14	15	16	17	18	19	20
21	22	23	24	25	26	27	28	29	30

When you subtract 10, the tens digit goes down 1.

The ones digit is the same.

So, $16 - 10 = 6$.

1. How can you use place value to subtract $12 - 10$? Color the numbers on the chart.

1	2	3	4	5	6	7	8	9	10
11	12	13	14	15	16	17	18	19	20
21	22	23	24	25	26	27	28	29	30

Fluency Flash

What is the sum or difference? Use place value.

2. $3 + 10 =$ _____

1	2	3	4	5	6	7	8	9	10
11	12	13	14	15	16	17	18	19	20
21	22	23	24	25	26	27	28	29	30

3. $18 - 10 =$ _____

1	2	3	4	5	6	7	8	9	10
11	12	13	14	15	16	17	18	19	20
21	22	23	24	25	26	27	28	29	30

Fluency Check

What is the sum or difference? Write the number.

4. $17 - 10 =$ _____

5. $4 + 6 =$ _____

6. $9 + 10 =$ _____

7. $15 - 10 =$ _____

8. $19 - 10 =$ _____

9. $8 - 5 =$ _____

10. $4 + 5 =$ _____

11. $8 + 10 =$ _____

12. $9 - 4 =$ _____

13. $2 + 3 =$ _____

14. $7 - 3 =$ _____

15. $2 + 10 =$ _____

Fluency Talk

How do you know that $5 + 10 = 15$?

How can you use doubles to subtract $5 - 3$?
Explain your work.

Glossary/Glosario

English	Spanish/Español

Aa

add (adding, addition) To join together sets to find the total or sum.

$$2 + 5 = 7$$

sumar (adición) Unir conjuntos para hallar el total o la suma.

$$2 + 5 = 7$$

---

addend Any numbers or quantities being added together.

$$2 + 3$$

2 is an addend and
  3 is an addend

sumando Números o cantidades que se suman.

$$2 + 3$$

2 es un sumando y
  3 es un sumando

---

analog clock A clock that has an hour hand and a minute hand.

minute hand

hour hand

reloj analógico Reloj que tiene manecilla horaria y minutero.

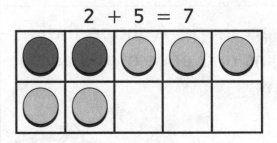

minutero

manecilla horaria

Cc

circle A closed, round figure. 	**círculo** Figura redonda y cerrarda. 
closed shape A shape that begins and ends at the same point. 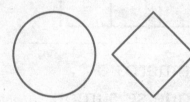	**figura cerrada** Figura que comienza y termina en el mismo punto. 
column A column goes up and down on a number chart.	**columna** Una columna sube y baja en una tabla numérica.
compare To look at objects, shapes, or numbers and see how they are alike or different.	**comparar** Observar objetos, formas o números para saber en qué se parecen y en qué se diferencian.

composite shape A figure made up of two or more shapes.	**figura compuesta** Figura formada por dos o más figuras.

 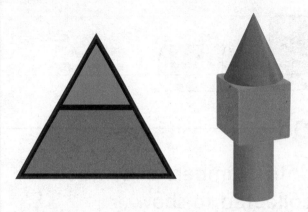

cone A 3-dimensional shape with 1 round base and 1 apex.	**cono** Figura tridimensional con una base redonda y un punto de unión.

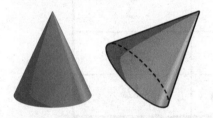 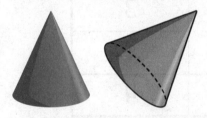

cube A 3-dimensional shape with 6 square faces, 8 vertices, and 12 edges.	**cubo** Figura tridimensional con 6 caras cuadradas, 8 vértices y 12 bordes.

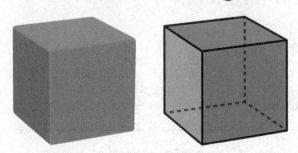 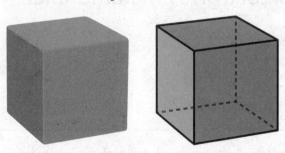

cylinder A 3-dimensional shape with 2 round faces and a curved surface.

cilindro Figura tridimensional con 2 caras redondas y una cara curva.

Dd

data Numbers or symbols collected to show information.

Name	Number of Pets
Mary	3
James	1
Alonzo	4

datos Números o símbolos que se reúnen para mostrar información.

Nombre	Número de mascotas
Mary	3
James	1
Alonzo	4

difference Subtracting one number from another number gives the difference.

$$3 - 1 = 2$$

↑

The difference is 2.

diferencia Restando un número de otro número da la diferencia.

$$3 - 1 = 2$$

↑

La diferencia es 2.

digit A symbol used to write numbers. The ten digits are: 0, 1, 2, 3, 4, 5, 6, 7, 8, 9.

dígito Símbolo usado para escribir números. Los diez dígitos son: 0, 1, 2, 3, 4, 5, 6, 7, 8, 9.

doubles Two addends that are the same number.

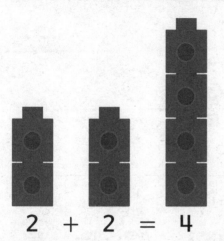

2 + 2 = 4

dobles Dos sumandos que son el mismo número.

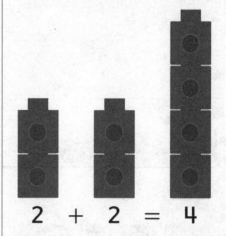

2 + 2 = 4

Ee

edge The line where two sides or faces meet.

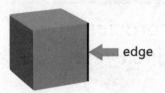

borde Línea donde dos lados o caras se unen.

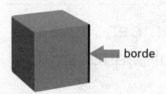

equal shares Each share is the same size.

Example: This muffin is cut into 2 equal shares.

partes iguales Cada una de las partes tiene el mismo tamaño.

Ejemplo: Este pastelillo está cortado en 2 partes iguales.

English	Spanish/Español
equal sign (=) Having the same value as or is the same as. 2 + 4 = 6 equal sign	**signo igual** (=) Que tienen el mismo valor o son lo mismo. 2 + 4 = 6 signo igual
equal to (=)  6 = 6 6 is equal to or the same as 6	**igual a** (=)  6 = 6 6 es igual o lo mismo que 6
equation A number sentence that includes an equal sign. 5 + 7 = 12	**ecuación** Una oración numérica que incluye el signo igual. 5 + 7 = 12

Ff

face The flat part of a 3-dimensional figure. Example: A square is a face of a cube. 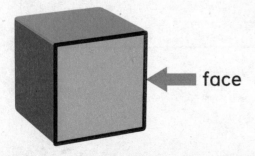 face	**cara** La parte plana de una figura tridimensional. Ejemplo: Un cuadrado es la cara de un cubo.  cara

English	Spanish/Español
fact family Addition and subtraction sentences that use the same numbers. Sometimes called *related facts*.	**familia de operaciones** Enunciados de suma y resta que tienen los mismos números. Algunas veces se llaman *operaciones relacionadas*.

$$6 + 7 = 13 \quad 13 - 7 = 6$$
$$7 + 6 = 13 \quad 13 - 6 = 7$$

$6 + 7 = 13 \quad 13 - 7 = 6$ $7 + 6 = 13 \quad 13 - 6 = 7$	$6 + 7 = 13 \quad 13 - 7 = 6$ $7 + 6 = 13 \quad 13 - 6 = 7$
fewer Not as many or a smaller amount.	**menos** El número o grupo con menos.

There are fewer yellow counters than red ones.

Hay menos fichas amarillas que fichas rojas.

fourths Four equal parts of a whole. Each part is a fourth, or a quarter, of the whole.	**cuartos** Cuarto partes iguales de un todo. Cada parte es un cuarto, o la cuarta parte del todo.

Gg

greater than (>) When an amount is larger than another amount.

Example: 7 is greater than 2 or 7 > 2.

mayor que (>) Cuando una cantidad es más grande que otro.

Ejemplo: 7 es mayor que 2 o 7 > 2.

Hh

half hour (or half past) One half of an hour is 30 minutes. Sometimes called *half past* or *half past the hour.*

media hora (o y media) Media hora son 30 minutos. A veces se dice hora y media.

English	Spanish/Español
halves Two equal parts of a whole. Each part is a half of the whole.	**mitades** Dos partes iguales de un todo. Cada parte es la mitad de un todo.
hexagon A 2-dimensional shape that has 6 sides. 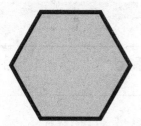	**hexágono** Una figura bidimensional con 6 lados. 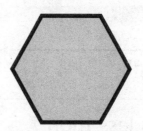
hour A unit of time. I hour = 60 minutes  	**hora** Unidad de tiempo. I hora = 60 minutos  
hour hand The hand on a clock that tells the hour. It is the shorter hand.  **hour hand**	**manecilla horaria** Manecilla del reloj que indica la hora. Es la manecilla más corta.  **manecilla horaria**

LI

length How long or how far away something is.

longitud La mayor de las dos dimensiones principales que tienen las cosas o figuras planas.

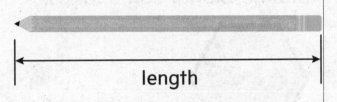

length

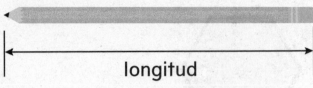

longitud

less than (<) 4 is less than 7.

menor que (<) 4 es menor que 7.

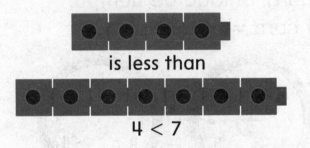

is less than

4 < 7

es menor que

4 < 7

English	Spanish/Español
long (longer, longest) A way to compare the lengths of objects.	**largo (más largo, el más largo)** Forma de comparar la l ongitud de objetos.

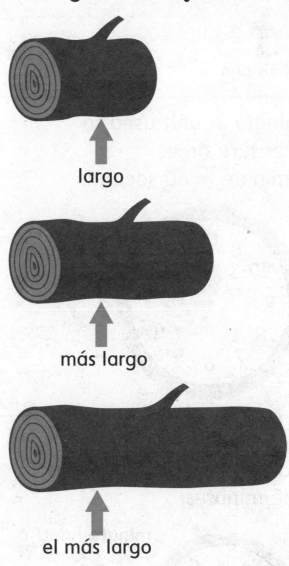

long

longer

longest

largo

más largo

el más largo

Mm

measure To find the length, height, or weight using standard or nonstandard units.

medir Hallar la longitud, estatura o peso mediante unidades estándar o no estándar.

English	Spanish/Español
minus (–) The sign used to show subtraction. $$5 - 2 = 3$$ minus sign	**menos (–)** Signo que indica resta. $$5 - 2 = 3$$ signo menos
minute A unit used to measure time. I minute = 60 seconds 	**minuto** Unidad para medir tiempo. I minuto = 60 segundos 
minute hand The longer hand on a clock that tells the minutes.  minute hand	**minutero** La manecilla más larga del reloj que indica los minutos.  minutero

more A larger or greater amount.

más Una cantidad mayor o más grande.

There are more red counters than yellow ones.

Hay más fichas rojas que amarillas.

Nn

number line A line with number labels.

recta numérica Recta con marcas de números.

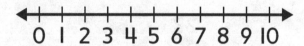

0 1 2 3 4 5 6 7 8 9 10

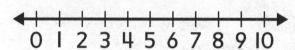

0 1 2 3 4 5 6 7 8 9 10

Oo

one-fourth One of 4 equal shares.

un cuarto Una de 4 partes iguales.

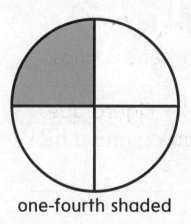

one-fourth shaded

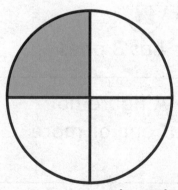

un cuarto sombreado

English	Spanish/Español

one-half One of 2 equal shares.

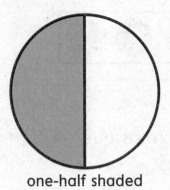

one-half shaded

mitad Una de 2 figuras iguales.

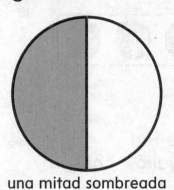

una mitad sombreada

---

ones The numbers in the range of 0–9. It is the place value of a number.

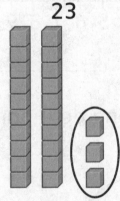

This number has 3 ones.

unidades Los números en el rango de 0 a 9. Es el valor posicional de un número.

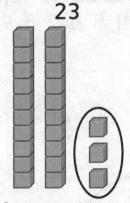

Este número tiene 3 unos.

---

open shape A figure not connected at one or more points.

figura abierta Figura que no está unida en uno o más puntos.

Pp

pattern An order that a set of objects or numbers follows over and over.

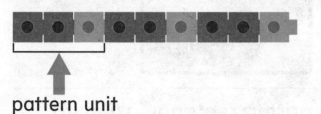

pattern unit

patrón Orden que sigue continuamente un conjunto de objectos o números.

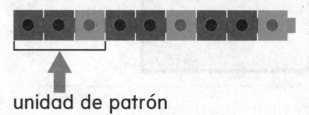

unidad de patrón

picture graph A graph that has different pictures to show information collected.

gráfica con imágenes Gráfica que tiene diferentes imágenes para ilustrar la información recopilada.

place value The value given to a digit by its place in a number.

65
6 tens
5 ones

valor posicional Valor de un dígito según el lugar en el número.

65
6 decenas
5 unidades

plus (+) The sign used to show addition.

$$4 + 5 = 9$$

plus sign

más (+) Símbolo para mostrar la suma.

$$4 + 5 = 9$$

signo más

Rr

rectangle A shape with 4 straight sides and 4 vertices.

rectángulo Figura con 4 lados y 4 esquinas.

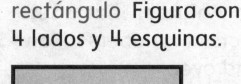

rectangular prism A 3-dimensional shape with 6 faces, 8 vertices, and 12 edges.

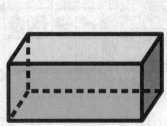

prisma rectangula Figura tridimensional con 6 caras, 8 esquinas y 12 bordes.

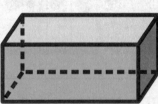

related fact(s) Basic facts using the same numbers. Sometimes called a *fact family*.

6 + 4 = 10   10 − 6 = 4
4 + 6 = 10   10 − 4 = 6

operaciones relacionadas Operaciones básicas en las cuales se usan los mismos números. También se llaman familias de operaciones.

6 + 4 = 10   10 − 6 = 4
4 + 6 = 10   10 − 4 = 6

row A row goes left to right on a number chart.

fila Una fila se lee de izquierda a derecha en una tabla numérica.

Ss

short (shorter, shortest) To compare length or height of two (or more) objects.

corto (más corto, el más corto) Comparar la longitud o la altura de dos (o más) objetos.

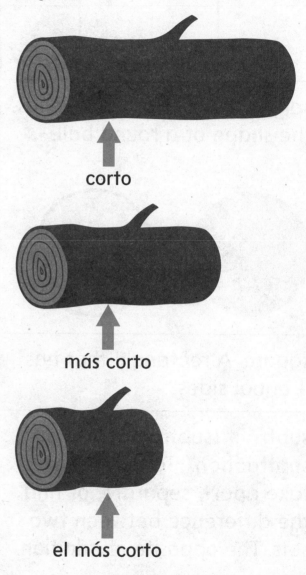

short

shorter

shortest

corto

más corto

el más corto

English	Spanish/Español
side One of the lines that make up a shape.	**lado** Uno de la lãneas que compone una figura.

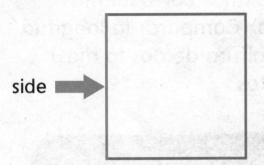 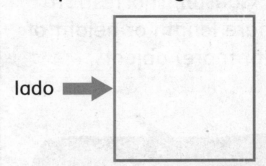

sphere A solid shape that has the shape of a round ball.	**esfera** Figura sólida con la forma de una pelota redonda.

square A rectangle that has 4 equal sides.	**cuadrado** Rectángulo que tiene 4 lados iguales.

subtract (subtracting, subtraction) To take away, take apart, separate, or find the difference between two sets. The opposite of addition.	**restar (resta, sustracción** Eliminar, quitar, separar o hallar la diferencia entre dos conjuntos. Lo opuesto de la suma.

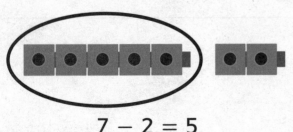 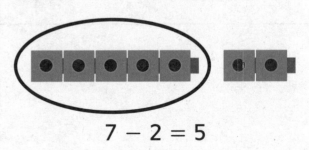

$$7 - 2 = 5$$ $$7 - 2 = 5$$

sum Adding two or more numbers gives the sum. $$2 + 4 = 6$$ ↑ sum	**suma** Sumando dos o más números da la suma. $$2 + 4 = 6$$ ↑ suma

Tt

tally mark(s) A mark used to record data collected in a survey. ⪢⪢⪢ ‖	**marca(s)** Símbolo usado para anotar datos de una encuesta. ⪢⪢⪢ ‖
teen number A number that is 1 group of ten and some ones. Example: 12	**números del 11 al 19** Un número formado por una decena y algunas unidades. Ejemplo: 12
ten One group of 10 ones is 1 ten. 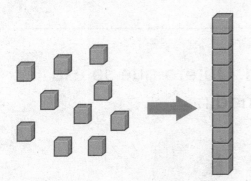	**decena** Un grupo de 10 es una decena. 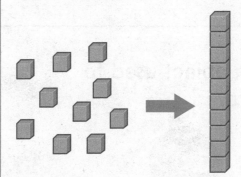

English	Spanish/Español
tens A place value of a number. Example: In the number 23, the 2 is in the tens place.	**decenas** El valor posicional de un número. Ejemplo: En el número 23, el 2 está en el lugar de las decenas.

23

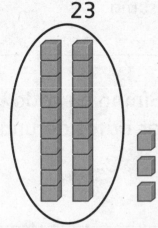

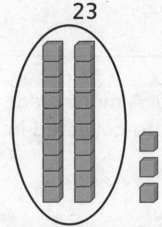

This number has 2 tens.	Este número tiene 2 decenos.

| **triangle** A shape with 3 sides. | **trapecio** Figura con 3 lados. |

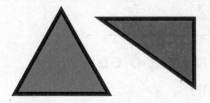

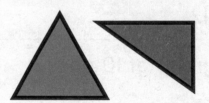

Uu

| **unit** An object used to measure. | **unidad** Objeto que se usa para medir. |

English	Spanish/Español
unknown A missing number in an equation.	**incógnita** El número que falta en una ecuación.
$9 + ? = 10$	$9 + ? = 10$

Ww

whole The entire object.	**el todo** El objeto completo.

  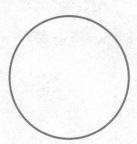